BARRON'S

VISUAL LEARNING

Biology

T0004075

Copyright © UniPress Books Limited 2021

Published by arrangement with UniPress Books Ltd

Publisher: Nigel Browning

Page design and illustrations: Lindsey Johns

Project manager: Kate Duffy

Editorial consultant: Cynthia Pfirrmann

First edition published in North America by Kaplan, Inc.,
d/b/a Barron's Educational Series

All rights reserved. No part of this book may be reproduced
in any form or by any means without the written permission
of the copyright owner.

Published by Kaplan North America, LLC, d/b/a Barron's Educational Series

1515 West Cypress Creek Road
Fort Lauderdale, FL 33309

www.barronseduc.com

ISBN: 978-1-5062-6761-6

Kaplan North America, LLC, d/b/a Barron's Educational Series print books are
available at special quantity discounts to use for sales promotions,
employee premiums, or educational purposes. For more information
or to purchase books, please call the Simon & Schuster
special sales department at 866-506-1949.

Printed in China

10 9 8 7 6 5 4 3 2

Dr. Helen Pilcher is a science writer, presenter, and performer.
Her previous books include *Life Changing: How Humans are Altering
Life on Earth*, *Bring Back the King: The New Science of De-extinction*, and
Mind Maps Biology. She writes for *Nature*, *New Scientist*, and *Science Focus*.
Dr. Pilcher has a Ph.D. in Cell Biology from London's Institute of
Psychiatry, and ran the Royal Society's Science in Society Program
before becoming a full-time writer. She regularly gives
science talks to schools and at festivals.

BARRON'S

VISUAL LEARNING

Biology

AN ILLUSTRATED GUIDE FOR ALL AGES

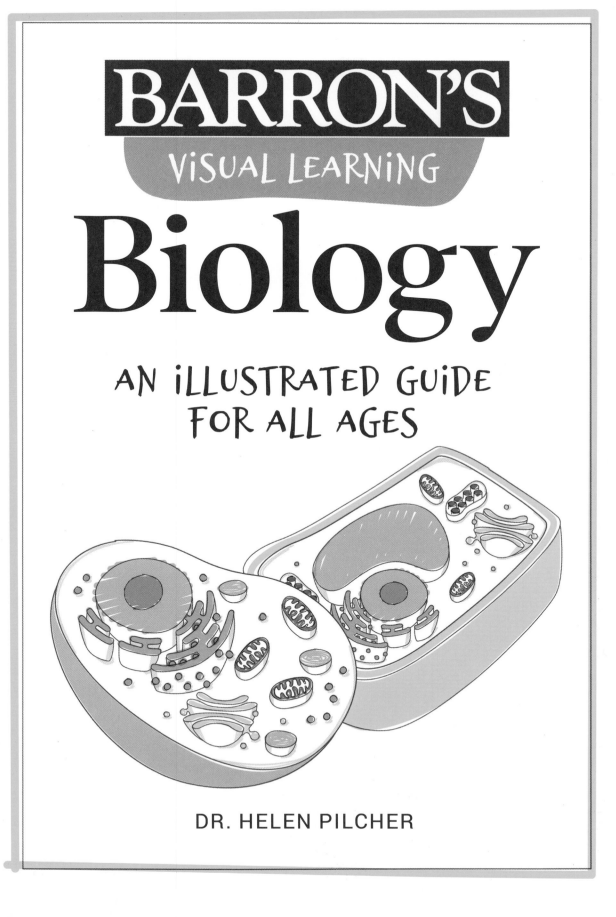

DR. HELEN PILCHER

CONTENTS

Introduction 6

1 Biology Basics 8

What Is Life? 9

The Chemistry of Life 10

The Molecules of Life 12

The Study of Biology 15

RECAP 16

2 Cells 18

Cell Basics 19

Microscopy 20

Cell Structure 22

Eukaryotic versus Prokaryotic Cells ... 25

Cell Division 26

Cell Transport 28

Stem Cells and Specialization 30

Cell Organization 32

RECAP 34

3 Genetics 36

DNA 37

Genetic Inheritance 40

Reproduction 42

Sex Determination 44

Mutations 46

Gene Editing 48

Nature and Nurture 50

RECAP 52

4 Evolution 54

Charles Darwin and the Voyage
of the *Beagle* 55

Evolution by Natural Selection ... 56

Extinction 59

Evidence for Evolution 60

Human Evolution 66

RECAP 68

5 Organizing Life 70

The Importance of Classification ... 71

Classification 72

Prokaryotes 74

Eukaryotes 76

Classifying Animals 83

Viruses 86

RECAP 88

6 Metabolism — 90

Chemical Reactions and Pathways — 91

Enzymes — 92

Metabolic Rate — 94

Control of Metabolism — 96

Respiration — 98

Photosynthesis — 100

RECAP — 102

7 Plant Structure and Function — 104

Transpiration — 105

The Vascular System — 106

Plant Growth — 107

Leaf Structure — 108

Plant Hormones — 110

Plant Deficiencies, Diseases, and Defenses — 112

RECAP — 114

8 Human Structure and Function — 116

Human Organ Systems — 117

Homeostasis — 118

Human Nervous System — 120

Human Brain — 122

Sense Organs — 124

Endocrine System — 128

Human Digestive System — 130

Circulatory and Respiratory Systems — 132

Skeletal and Muscular Systems — 134

RECAP — 136

9 Human Health and Disease — 138

Health Inequalities — 139

Communicable Diseases — 140

Noncommunicable Diseases — 144

Drugs and Diseases — 148

Lifestyle and Health — 150

RECAP — 152

10 Ecology — 154

Ecosystems — 155

Interdependence — 156

Environmental Influences on Life — 158

Adaptations — 160

Competition — 162

Keystone Species — 165

Global Recycling of Elements — 166

RECAP — 170

11 Biology in the Twenty-First Century — 172

The Anthropocene — 173

Planetary Boundaries — 174

The Greenhouse Effect — 176

Climate Change — 178

Biodiversity Loss and Extinction — 182

RECAP — 186

Index — 188

Acknowledgments — 192

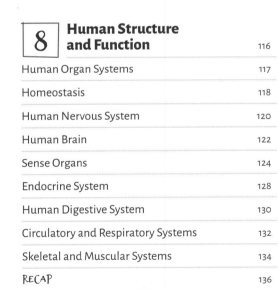

INTRODUCTION

This book is all about biology—the study of life. From the driest of deserts to the most luscious of rain forests, from the depths of the oceans to the far reaches of the atmosphere, our planet is teeming with life. It exists in a bountiful, diverse collection of forms, and by studying life, we can understand more about the planet that we live on and more about ourselves.

In this extensively illustrated book, we'll be exploring many different areas of biology in an attempt to stimulate curiosity and satisfy a thirst for knowledge. Did you know, for example, that there is an all-female species of lizard that can reproduce without the need for males? Well, you do now, and there are plenty more gems just like that waiting for you in the pages that follow.

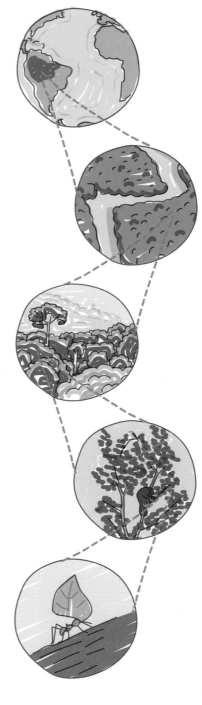

Visual Learning Biology is written for anyone with an interest in biology and the world around them. Complex topics are distilled with clarity, and scientific terms are carefully explained. This book is written with the assumption that anyone can understand just about anything, as long as it is presented in an appropriate and approachable style.

We all experience the world in different ways, and we all have different learning styles. Some people benefit in particular from visual learning. These learners remember things better when information is presented in a visual way. This book is particularly aimed at those with a visual learning style. Where possible, facts and ideas are distilled into pictures and diagrams. Lengthy

Some species, such as this whiptail lizard, can reproduce asexually.

The Earth is full of many different ecosystems.

chunks of text have been replaced with diagrams and flow charts. The illustrations are eye-catching and colorful, and the legends are informative and succinct.

Visual Learning Biology is deliberately broad in scope. It covers many of the core topics that are covered in biology courses, but it also includes some cutting-edge developments that have yet to make it to the curriculum. Did you know, for example, that scientists are seeking to bring the woolly mammoth back from extinction? Or that identical twins end up different, in part, because of variations in gene activity? These so-called epigenetic changes are now helping to shed light on some human diseases, and helping scientists to devise new therapies and treatments.

The book is divided into 11 chapters. It begins by asking the most fundamental question of all—what is life?—and then continues by investigating the building blocks of life, in the form of structures, such as DNA,

Charles Darwin proposed the theory of evolution by natural selection.

proteins, and cells. It explores the diversity of living things and how they are categorized and delves into what is arguably the greatest scientific theory of all time, the theory of evolution by natural selection. You'll learn how life changes over time and how new species arise, as well as the many diverse strands of evidence that support this theory.

You'll also learn about the workings of the human body and what happens when cells stop working properly and diseases emerge. Next, you'll explore plant life and the highly specialized features that enable plants to harness solar power in order to make food.

Then, in the final chapters, *Visual Learning Biology* turns to the broader natural world and examines the interconnectedness of all living things. As we lurch into a time of climate crisis, this book also highlights some of the many dangers that now threaten the stability of our planet. Finishing on a positive note, it then explores the many ways that governments, communities, and individuals can make a difference to help preserve the biodiversity and biology of the future. Welcome to *Visual Learning Biology*!

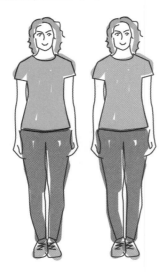

Identical twins contain DNA that is almost identical, but can develop into very different people.

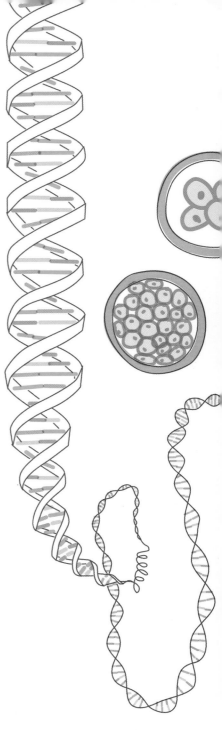

Cells contain information in the form of DNA: a double-stranded, helical structure made of complementary base pairs.

CHAPTER 1

BIOLOGY BASICS

Our planet is covered in life. Life comes in many shapes and sizes, from the tiniest bacteria to the tallest tree, and to the biggest blue whale. Biology is the study of living things and it is a remarkably diverse field. The word "biology" is derived from Greek. *Bios* means "life" and *logos* means "study."

But what exactly is life, and what are living things made of? In this chapter, you'll explore the basics of life and biology.

WHAT iS LiFE?

Living things, such as people and plants, are made from the same chemicals as nonliving things, such as rocks and water, yet living and nonliving things are very different. Although it's easy to point out something that's alive, it's actually difficult to define what life is. Some definitions explain what life does, rather than what life is. To pin down the essence of life, biologists have assembled a list of the properties that are unique to living organisms.

Properties of living organisms

Organization: Living things are made up of cells that contain the chemical deoxyribonucleic acid (called DNA).

Growth: Living things develop and grow. Complex life-forms start as a single cell, which then divides repeatedly to form an adult individual.

Reproduction: Living things create offspring and pass their DNA on to the next generation.

Sensitivity: Living things detect and respond to stimuli, e.g. hawks spot rabbits, and rabbits are watchful of hawks.

Respiration: Chemical reactions help living things to break down nutrients to release energy.

Nutrition: Living things take in nutrients, such as organic molecules and mineral ions, and they use them as a source of energy, e.g. rabbits eat grass.

Excretion: Living things excrete waste products and toxic substances.

Movement: Living things can move and change their position, e.g. rabbits run away from predators. Plants turn toward the sun.

THE CHEMISTRY OF LIFE

All life is made from the same tiny chemical units: atoms and elements. Chemical substances, such as carbon, oxygen, and hydrogen, are called **elements**. **Atoms** are the smallest units of any particular element. Although the amounts of the different atoms vary from one element to another, all life on Earth shares this same basic level of chemistry.

Atoms have a central nucleus made of protons and neutrons, surrounded by rings of electrons. The protons and electrons are attracted to each other. This holds the atom together.

Protons are subatomic particles found in the nucleus. They are positively charged.

Neutrons are subatomic particles found in the nucleus. They have no charge.

Electrons are subatomic particles found in the rings that surround an atomic nucleus. They are negatively charged.

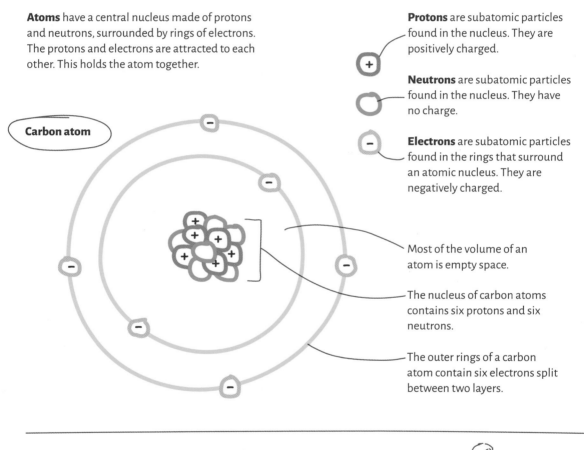

Carbon atom

Most of the volume of an atom is empty space.

The nucleus of carbon atoms contains six protons and six neutrons.

The outer rings of a carbon atom contain six electrons split between two layers.

Living things contain vast numbers of atoms. The adult human body, for example, contains around 7×10^{27} atoms. That is the number 7, followed by 27 zeros.

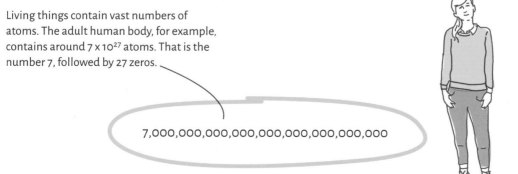

7,000,000,000,000,000,000,000,000,000

Atoms in living organisms

All living things contain four common elements: oxygen, carbon, hydrogen, and nitrogen. Along with calcium and phosphorus, these elements make up over 99% of our mass. Most of the remainder is potassium, sulfur, sodium, chlorine, and magnesium. These 11 elements are called **major elements** because they are needed for life.

Trace elements, such as iron, manganese, and zinc, are also needed, but in very small or "trace" amounts. For example, iron is a trace element. It is needed by all species. In mammals, iron forms part of a bigger molecule called hemoglobin. Hemoglobin helps to move oxygen around our bodies.

Copper is another trace element. Around 100 years ago, scientists realized its importance after finding that rats with too little copper in their diets struggled to make red blood cells. We now know that copper helps the body to use iron. It also does many other things, including helping the body to fight infection and combining with proteins to make important enzymes.

All in all, humans need 25 elements in order to live. Plants, in contrast, need just 17 elements in order to survive.

There are about 92 naturally occurring elements. Some of these are toxic to living things. Arsenic is an element that occurs naturally in some rocks and soil. Sometimes arsenic dissolves out of these natural reserves and winds up in the water supply. This is a big problem because arsenic can be fatal.

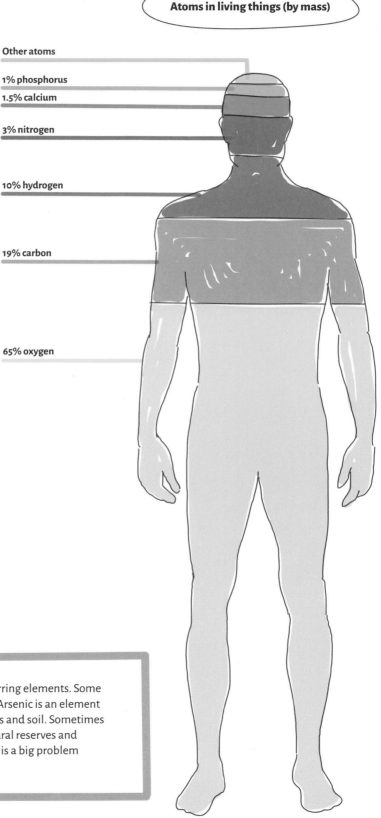

Atoms in living things (by mass)

Other atoms

1% phosphorus

1.5% calcium

3% nitrogen

10% hydrogen

19% carbon

65% oxygen

THE MOLECULES OF LiFE

Atoms and elements are organized into larger structures called molecules. A **molecule** is any substance made from two or more atoms that are chemically bonded together. Life is made of biological molecules or "biomolecules."

Biomolecules are any of the many different molecules that are made by cells and living things. They perform many different vital functions. There are four major different types: carbohydrates, lipids, nucleic acids, and proteins. (Nucleic acids, such as DNA and RNA, are covered in Chapter 3.)

For example, glucose is a molecule. Cells use glucose as a form of energy. Each molecule of glucose is made up of 6 carbon atoms, 12 hydrogen atoms, and 6 oxygen atoms. Carbon has the chemical symbol C. Hydrogen has the chemical symbol H, and oxygen is shortened to O. So, the molecular formula for glucose is $C_6H_{12}O_6$.

Glucose's molecular structure

$C_6H_{12}O_6$

- Carbon
- O Oxygen
- H Hydrogen
- O H Hydroxide molecule

Carbohydrates are based on simple sugars.

Carbohydrates

Carbohydrates are a large group of biomolecules that are found in food and in living tissues. Carbohydrate molecules contain carbon, hydrogen, and oxygen. These elements are arranged to form sugar molecules. The sugars are arranged into chains, and the chains come in different lengths.

There are two basic types of carbohydrates. Simple carbohydrates are small molecules. Complex carbohydrates are made of long chains of simple sugar units that are chemically bonded together.

Carbohydrate molecules

Glucose is a simple carbohydrate. It contains one sugar unit.

Sucrose is a simple carbohydrate. It is made of two different sugar units joined together. Sucrose is the sugar that we put in our tea or coffee.

Starch is a complex carbohydrate.

Carbohydrates are an important source of energy. They provide cells with the fuel that they need to carry out vital chemical reactions. Many of the fruit and vegetables we eat are made up of simple carbohydrates.

Good sources of complex carbohydrates include potatoes, bread, and pasta. Carbohydrates are also used as a building material. Plant cell walls are made of a complex carbohydrate called cellulose.

Lipids

Lipids are substances, such as fats, which are solid at room temperature, and oils, which are liquid at room temperature. Lipid biomolecules contain carbon, hydrogen, and oxygen atoms. They are insoluble in water.

Lipids are made up of two basic components: glycerol and fatty acids. Three fatty acids are joined to each molecule of glycerol. Different lipids contain different types of fatty acid. They determine whether the lipid will be a liquid oil or a solid fat.

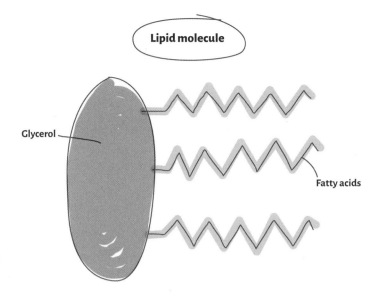

Lipid molecule

Glycerol

Fatty acids

Lipids are an important store and source of energy. Foods such as butter, cheese, nuts, seeds, and fish contain lipids. Once you have used up all of your glucose stores, your body will eventually start to break down your fat.

Mammals store their long-term food reserves in fat cells called **adipose** cells. Adipose tissue also helps to insulate the body and forms a protective, cushioning layer around vital organs such as the heart.

Lipids are important structural molecules. **Phospholipids** are a type of lipid. They help to form the outer cell membrane of animal cells.

Phospholipids in a cell membrane

Water-attracting heads assemble on the outside of the bilayer structure.

Channels in the membrane help important molecules to pass in and out of the cell.

If there is water present, phospholipids naturally line up to form a double-layered (bilayer) structure.

Water-repelling tails are tucked away on the inside of the bilayer structure.

Steroids are another type of lipid. They are an important component of cell membranes, and they also act as signaling molecules. Cholesterol and testosterone are both steroids.

Cell membrane

Cell

Proteins

Proteins are large, complicated molecules made from smaller molecules called **amino acids**. Special bonds link the amino acids together to form long chains.

Different proteins contain various amino acids arranged in different orders. Proteins contain carbon, hydrogen, oxygen, and nitrogen.

Around 15% of your body mass is made of protein. Protein-dense foods include meat, cheese, fish, and dried beans, such as chickpeas and lentils.

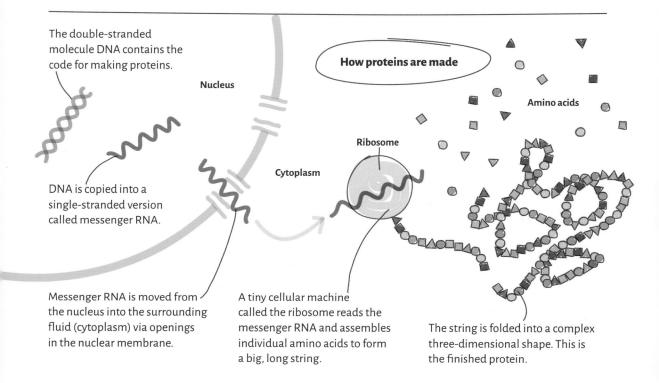

The double-stranded molecule DNA contains the code for making proteins.

Nucleus

How proteins are made

Amino acids

Ribosome

Cytoplasm

DNA is copied into a single-stranded version called messenger RNA.

Messenger RNA is moved from the nucleus into the surrounding fluid (cytoplasm) via openings in the nuclear membrane.

A tiny cellular machine called the ribosome reads the messenger RNA and assembles individual amino acids to form a big, long string.

The string is folded into a complex three-dimensional shape. This is the finished protein.

FUNCTION OF PROTEINS

Proteins do a lot of important jobs. They have the following roles.

★ **Enzymes**: They speed up chemical reactions, e.g. digestive enzymes help to break down food.

★ **Hormones**: Insulin is a hormone released from the pancreas that prompts other tissues to take up glucose.

★ **Storage molecules**: The main protein in mammalian milk is casein, which stores amino acids, so baby mammals can use them.

★ **Antibodies**: Proteins help the immune system to detect and destroy disease-causing microbes.

★ **Transporters**: Proteins, such as hemoglobin, move substances around the body.

Casein in breast

Antibodies are produced by white blood cells, which circulate in the blood.

Digestive enzymes in stomach

Similarly, hemoglobin is found in red blood cells that circulate in the blood.

Insulin in pancreas

THE STUDY OF BIOLOGY

Some biologists study life at the molecular level. For example, they might study the atomic structure of a particular lipid or how certain amino acids are folded to make a particular protein. Biology, however, is bigger than this. Atoms and molecules form cells and larger organized structures, which combine to form organisms. The organisms live in groups in ecosystems, which combine to make up the natural world. Biologists study every aspect of this hierarchy.

Levels of biological study

Atoms: Tiny particles that make up all substances.

Molecules: A group of atoms bonded together.

Species: Groups of similar individuals that can interbreed to produce fertile offspring. The groups may be separated geographically.

Organelles: Tiny structures found inside cells that carry out specialized functions.

Ecosystem: Communities of living things and the physical environment in which they live.

Cells: Membrane-bound units containing fundamental biomolecules, such as proteins, lipids, and DNA.

Population: Groups of similar individuals that live close to one another and interbreed.

Tissues: Groups of cells with a similar structure and function. They work together to do a specific job.

Biosphere: Parts of the Earth's surface and atmosphere that are occupied by living things.

Organs: Structures made from different tissue types that work together to perform a particular function.

System: Complicated structures containing multiple organs, working together to perform a specific function.

Organism: An individual animal, plant, fungus, or other living thing.

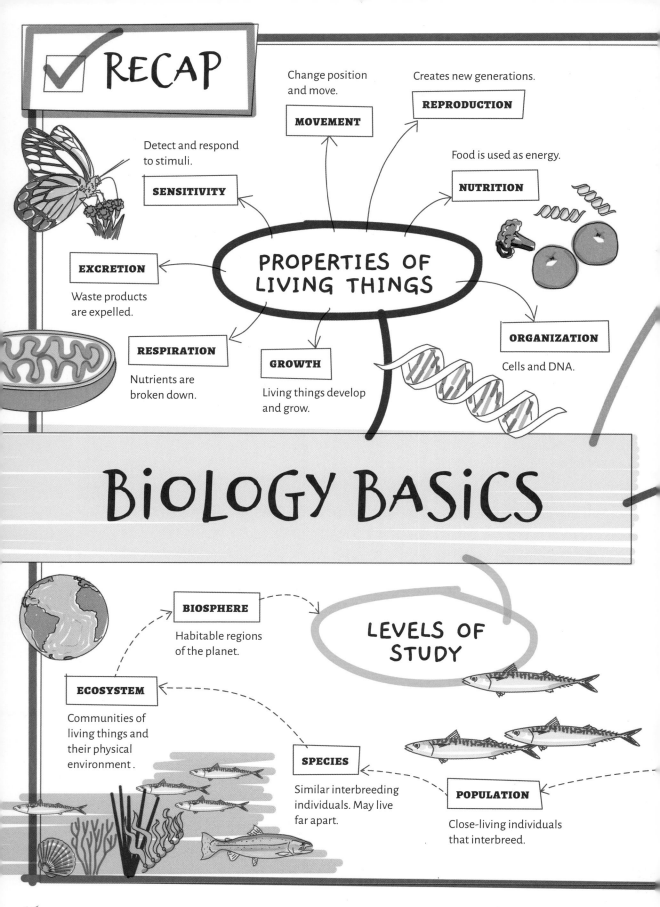

Change position
and move.

Creates new generations.

MOVEMENT

REPRODUCTION

Detect and respond
to stimuli.

Food is used as energy.

SENSITIVITY

NUTRITION

PROPERTIES OF LIVING THINGS

EXCRETION

Waste products
are expelled.

ORGANIZATION

Cells and DNA.

RESPIRATION

Nutrients are
broken down.

GROWTH

Living things develop
and grow.

BiOLOGY BASiCS

BIOSPHERE

Habitable regions
of the planet.

LEVELS OF STUDY

ECOSYSTEM

Communities of
living things and
their physical
environment.

SPECIES

Similar interbreeding
individuals. May live
far apart.

POPULATION

Close-living individuals
that interbreed.

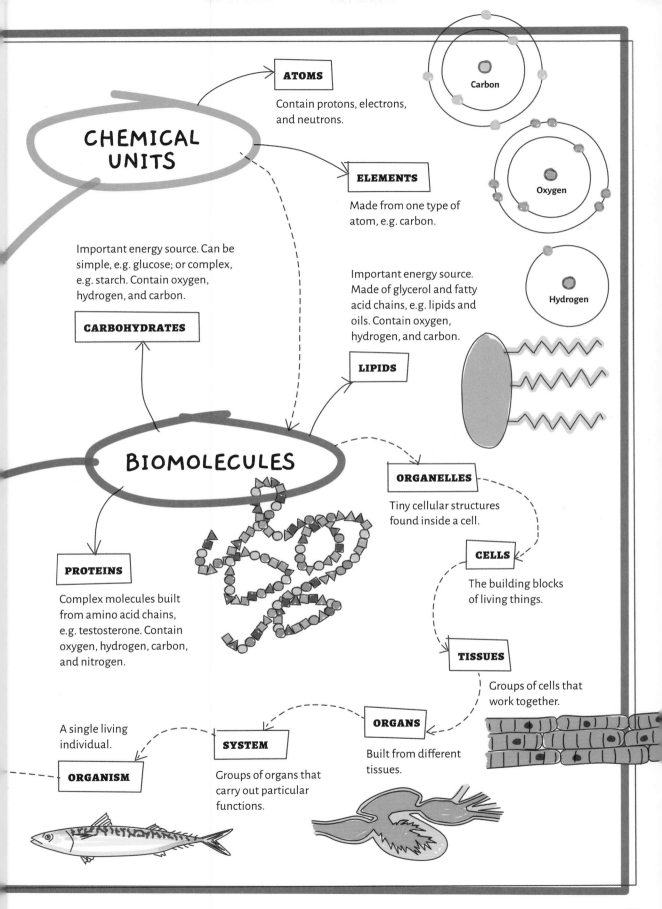

ATOMS

Contain protons, electrons, and neutrons.

Carbon

CHEMICAL UNITS

ELEMENTS

Made from one type of atom, e.g. carbon.

Oxygen

Hydrogen

Important energy source. Can be simple, e.g. glucose; or complex, e.g. starch. Contain oxygen, hydrogen, and carbon.

Important energy source. Made of glycerol and fatty acid chains, e.g. lipids and oils. Contain oxygen, hydrogen, and carbon.

CARBOHYDRATES

LIPIDS

BIOMOLECULES

ORGANELLES

Tiny cellular structures found inside a cell.

CELLS

The building blocks of living things.

PROTEINS

Complex molecules built from amino acid chains, e.g. testosterone. Contain oxygen, hydrogen, carbon, and nitrogen.

TISSUES

Groups of cells that work together.

A single living individual.

SYSTEM

ORGANS

Built from different tissues.

ORGANISM

Groups of organs that carry out particular functions.

CHAPTER 2

CELLS

All living things are made of cells. Cells are the basic building blocks of life. There are many different types of cells. The human body, for example, contains at least 200 different varieties, and each cell type is specialized to perform its own vital role.

Although they are small, cells are incredibly complex. They contain tiny structures called organelles, which act like microscopic machines and help the cell to function. You'll explore these structures and find out what goes on inside a cell.

CELL BASICS

Some living things, such as bacteria and amoebas, are relatively simple and made of just a single cell. Others, such as animals and plants, are more complex and made from a mix of different, specialized cells.

Unicellular life

Unicellular organisms are made from a single cell. There are different types of unicellular organisms. These include bacteria, certain fungi, and protozoans, such as amoebas (shown here).

Nucleus: DNA is stored here.

Cytoplasm: Important chemical reactions occur here.

Cell membrane: The flexible outer layer allows molecules to enter and leave the cell.

Pseudopodia: Tiny projections help the amoeba to move and eat.

Food particle: Pseudopodia help the amoeba to engulf food particles.

Multicellular life

Multicellular organisms are made from two or more cells. Complex multicellular organisms first appeared around 600 million years ago. They evolved when cells joined together and started to acquire new functions. The first multicellular animals would have been very simple—like a sea sponge. After that, multicellular organisms became more complex and evolved into the many different life-forms that inhabit our planet today.

A sea sponge (shown here) has a hollow body. It pumps water through its cavities in order to extract food. It contains a limited number of different cell types.

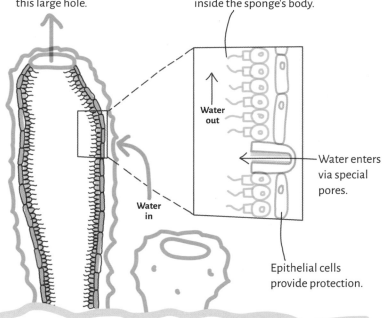

Water leaves via this large hole.

Collar cells beat their whiplike tails to create a water current inside the sponge's body.

Water out

Water enters via special pores.

Water in

Epithelial cells provide protection.

MICROSCOPY

Most cells are so small, they are impossible to see with the naked eye, so biologists use microscopes to study and photograph cells.

A **microscope** is an instrument that magnifies objects that are otherwise too small to be seen, such as many cells.

Cells come in different shapes and sizes. Human egg cells are just big enough to be seen with the naked eye.

Human egg cell

0.1 mm
(0.004 in)

Red blood cell

Red blood cells are too small to be seen with the naked eye.

0.008 mm
(0.0003 in)

Light microscopes

The first microscopes were **light microscopes**. These use visible light and lenses to magnify objects. Early light microscopes magnified objects a few hundred times. Today's light microscopes can magnify objects a few thousand times.

Light microscopes can be used to see cells and some of the big structures inside them, such as nuclei. Light microscopes are useful because they can be used to study living cells. This means that scientists can watch cells performing normal functions, such as dividing.

Coarse adjustment knob: This helps sharpen the image.

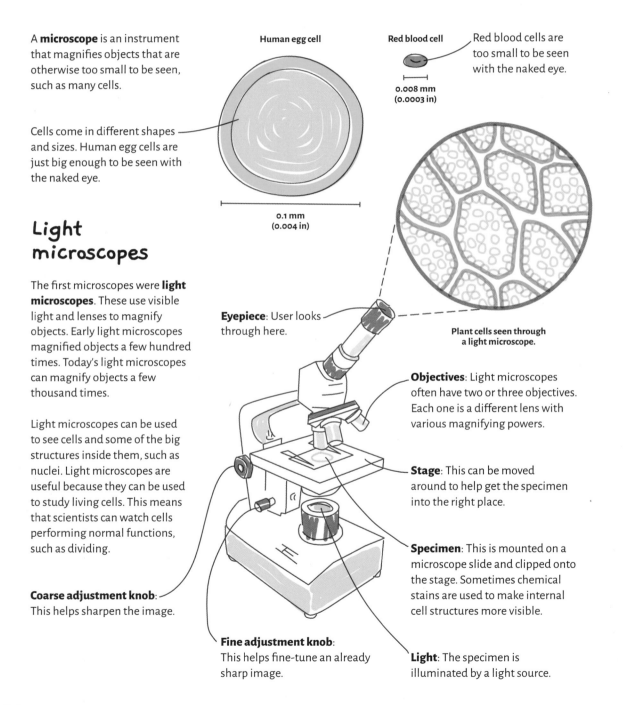

Plant cells seen through a light microscope.

Eyepiece: User looks through here.

Objectives: Light microscopes often have two or three objectives. Each one is a different lens with various magnifying powers.

Stage: This can be moved around to help get the specimen into the right place.

Specimen: This is mounted on a microscope slide and clipped onto the stage. Sometimes chemical stains are used to make internal cell structures more visible.

Fine adjustment knob: This helps fine-tune an already sharp image.

Light: The specimen is illuminated by a light source.

Electron microscopes

Electron microscopes use a beam of electrons to form an image. They were invented in the 1930s. They can magnify objects up to around 10 million times, so they can be used to study the minute structures found inside cells as well as other tiny structures, such as pollen grains.

Electron microscopes are large and expensive. Unlike light microscopes, they cannot be used to study living cells because the samples need to be kept in a vacuum. There are two main types of electron microscopy.

Scanning electron microscopy (SEM) uses a beam of electrons that moves back and forth over the surface of the specimen. It produces detailed 3D images.

Transmission electron microscopy (TEM) is performed on thin slices of a specimen. The beam of electrons passes through the slice. This can be used to produce detailed images of the structures inside cells.

Resolution is the ability to distinguish between two different points, so a higher resolution gives a sharper, more detailed image. Electron microscopes have a higher resolution than light microscopes.

Measuring the microscopic

Cells and the structures they contain are typically measured in millimeters (1 mm = 0.039 in), micrometers, and nanometers.

1 cm = 10 millimeters (mm)

1 mm = 1,000 micrometers (µm)

1 µm = 1,000 nanometers (nm)

A question of scale

Droplet = 2 mm

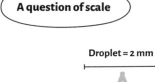

Bacterium = 2 µm

x 1,000

DNA = 2 nm

x 1,000

Magnification = Image size / Real size

So
Real size = Image size / Magnification

Therefore, if the magnification of the objective is x 40, and the image of the cell you are looking at is 1 mm (0.039 in), you can calculate the diameter of the cell.

The diameter of the cell = 1 mm / 40
= 0.025 mm or 25 µm.

CELL STRUCTURE

Plants and animals may be very different, but the cells they are made of share many common features. For example, their genetic material is enclosed in a nucleus, and their organelles sit in a gel-like substance called the cytoplasm. Bacterial cells are much smaller in comparison and their internal organelles are a little different. The genetic material floats freely in the cytoplasm and many of the standard animal and plant cell organelles are missing.

Animal cell

Animal cells are **eukaryotic**. This means they have a membrane-bound nucleus and a variety of organelles that carry out different functions.

The **cytoplasm** forms the bulk of the cell. It contains the nucleus and all the other structures that are found inside a cell. Many important events occur inside the cytoplasm, including cell division and key chemical reactions.

The **nucleus** is a small membrane-bound structure. It contains the vast majority of the cell's DNA.

The **cell membrane** is the outer layer that surrounds the cytoplasm. It controls the movement of substances in and out of the cell.

Mitochondria are tiny, battery-like structures found inside the cytoplasm. They provide the cell with energy, which they make by respiration. Mitochondria contain a tiny amount of the cell's DNA.

The **endoplasmic reticulum** is a system of flattened, tube-like structures. There are two types. Rough endoplasmic reticulum is covered in ribosomes and makes a lot of protein. Smooth endoplasmic reticulum has no ribosomes and is involved in making lipids.

Golgi bodies are relatively large organelles. They receive, make, modify, and distribute many different types of molecules, such as enzymes and other proteins.

Lysosomes are structures that help to break down big molecules into smaller ones.

Ribosomes are tiny protein-making factories. They link together amino acids to form proteins. Cells that are very biologically active, and make lots of proteins, have lots of ribosomes.

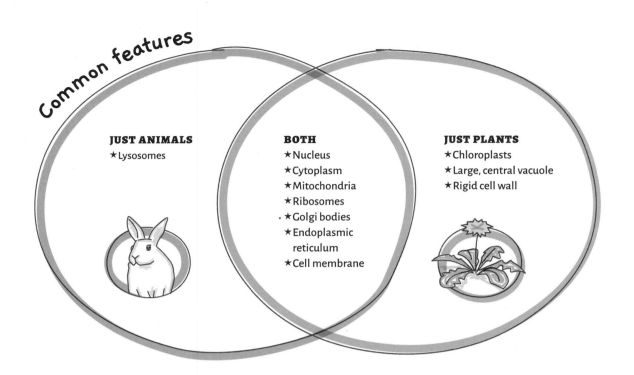

JUST ANIMALS
★ Lysosomes

BOTH
★ Nucleus
★ Cytoplasm
★ Mitochondria
★ Ribosomes
· ★ Golgi bodies
★ Endoplasmic
   reticulum
★ Cell membrane

JUST PLANTS
★ Chloroplasts
★ Large, central vacuole
★ Rigid cell wall

Plant cell

Plant cells are also eukaryotic. Just like animal cells, they carry the bulk of their DNA inside a membrane-bound nucleus and have specialized organelles to help them function. However, plants contain some structures that are not found in animal cells.

Plant and algae cells are surrounded by a thick, rigid **cell wall** that contains cellulose. This gives the cell strength and structure.

Chloroplasts are tiny structures packed with a green pigment called **chlorophyll**. This is the place where photosynthesis occurs. Chloroplasts can be found in all the green parts of a plant, such as the shoots and the leaves, but they are absent from the roots, which lack this green coloring.

There is a big, central space in the middle of the cytoplasm called a **vacuole**. The vacuole is filled with sap. This gives the cell structure and bulk and helps to keep the plant rigid.

Bacterial cell

Although they share a few common features, bacterial cells are very different from animal and plant cells. Bacterial cells are **prokaryotic**. This means that they have no large membrane-bound structures in their cytoplasm.

The **cell membrane** surrounds the cytoplasm. It is made from phospholipids. It helps control the movement of substances into and out of the cell.

The **cell wall** surrounds the cell membrane. This protective structure gives the bacterium a rigid, fixed shape. Bacterial cell walls are made of a molecule called peptidoglycan. Penicillin targets the peptidoglycan molecules inside bacterial cell walls. This causes the walls to break apart and kills the bacteria.

The **genetic material** floats freely inside the cytoplasm. This is a single loop of DNA.

The **cytoplasm** is a gel-like substance. Most of the cell's activities occur here.

Sometimes a **slime layer** surrounds the cell wall. This also helps to protect the bacterium.

Some bacteria are covered in hair-like structures called **pili**. They help the bacteria to stick to things and to move around.

Bacterial ribosomes assemble amino acid molecules together to make proteins. These ribosomes have a different structure from animal and plant ribosomes.

Sometimes bacteria contain small rings of DNA called **plasmids**.

Bacterial cells lack membrane-bound organelles such as nuclei and mitochondria.

Some bacteria have tails called **flagellae**. They thrash around and help the bacterium to move.

EUKARYOTIC VERSUS PROKARYOTIC CELLS

Life on Earth can be split into eukaryotic and prokaryotic forms. As we've already seen, eukaryotes contain large membrane-bound organelles, but prokaryotes do not.

Eukaryotic life forms are more specialized and advanced than prokaryotic ones. They can be multicellular, such as animals and plants, or unicellular, such as some algae and fungi. There are an estimated 9 million eukaryotic species on our planet.

Prokaryotic life-forms are less specialized. We are less familiar with these simple, small unicellular species.

Scientists think there could be up to a trillion different prokaryotic species on Earth.

Prokaryotes and some simple eukaryotes, such as amoebas, replicate by **binary fission**. Binary fission is a simple form of cell division.

Binary fission

The DNA loop and the plasmids replicate.

The cell gets bigger and the loops of DNA move to opposite ends of the cell.

The cytoplasm starts to divide and new cell walls form.

Two daughter cells are produced. They each contain one DNA loop and a variable number of plasmids. Binary fission enables prokaryotes to multiply very rapidly if the conditions are favorable. *E. coli* bacteria, for example, can replicate in as little as 20 minutes.

FEATURES OF EUKARYOTES AND PROKARYOTES

Feature	Eukaryotes Animal	Eukaryotes Plant	Prokaryotes
DNA	✓	✓	✓
Nucleus	✓	✓	✗
Cytoplasm	✓	✓	✓
Ribosomes	✓	✓	✓
Cell membrane	✓	✓	✓
Cell wall	✗	✓	✓
Mitochondria	✓	✓	✗
Golgi bodies	✓	✓	✗
Endoplasmic reticulum	✓	✓	✗
Chloroplasts	✗	✓	✗
Large sap-filled vacuoles	✗	✓	✗

CELL DIVISION

Sometimes cells in multicellular organisms need to divide; either to produce copies of themselves or to generate new, different cell types. Two types of cell division are involved: mitosis and meiosis. They occur when chunks of DNA in the nucleus, called chromosomes, are rearranged and new daughter cells are formed.

The cell cycle

Growth and DNA replication.

Mitosis: This is the stage of the cycle where cell division occurs.

Cell division by mitosis

Cell division by **mitosis** produces genetically identical daughter cells that are used to fuel growth and repair bodily damage. Mitosis occurs in cells all over the body. Dividing cells pass through a series of stages called the **cell cycle**. Mitosis, a form of asexual reproduction, is part of this cycle.

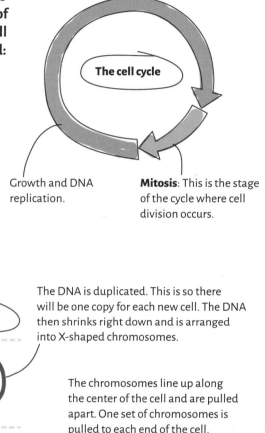

Mitosis

The DNA is duplicated. This is so there will be one copy for each new cell. The DNA then shrinks right down and is arranged into X-shaped chromosomes.

Growth and DNA replication

When a cell is not dividing, the DNA inside the nucleus is arranged in long tangled strings. In preparation for mitosis, the cell grows and makes extra copies of organelles, such as mitochondria and ribosomes.

Mitosis

The chromosomes line up along the center of the cell and are pulled apart. One set of chromosomes is pulled to each end of the cell.

Membranes form around the freshly separated chromosomes. These become the nuclei of the two new cells. The cytoplasm and cell membrane also divide.

Two new daughter cells are produced. They contain exactly the same DNA as the original parent cell, so they are genetically identical. In time, these daughter cells may divide by mitosis to produce more identical cells.

Cell division by meiosis

Cell division by **meiosis** produces sex cells or **gametes**, such as sperm, eggs, and spores. It involves two rounds of division. In humans, meiosis occurs in the ovaries and the testes. It is an important part of sexual reproduction.

Unlike other bodily cells, gametes only have one copy of each chromosome, rather than two. In humans, they have 23 single chromosomes as opposed to 23 pairs of chromosomes.

This is so that when two gametes combine, the resulting cell ends up with the correct amount. Cells divide by meiosis in order to produce gametes that have half the original number of chromosomes.

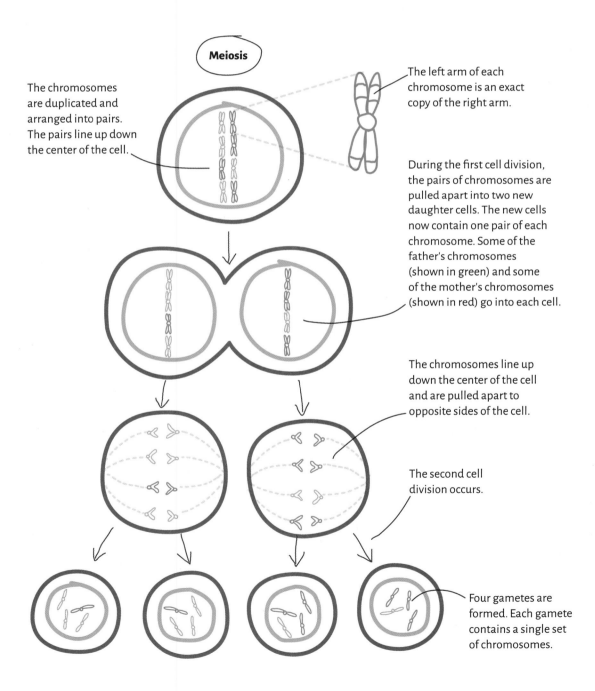

Meiosis

The chromosomes are duplicated and arranged into pairs. The pairs line up down the center of the cell.

The left arm of each chromosome is an exact copy of the right arm.

During the first cell division, the pairs of chromosomes are pulled apart into two new daughter cells. The new cells now contain one pair of each chromosome. Some of the father's chromosomes (shown in green) and some of the mother's chromosomes (shown in red) go into each cell.

The chromosomes line up down the center of the cell and are pulled apart to opposite sides of the cell.

The second cell division occurs.

Four gametes are formed. Each gamete contains a single set of chromosomes.

CELL TRANSPORT

Cells need to be able to control all the substances that enter and leave them. They need to be able to take in certain substances, such as nutrients and water, and dispose of others, such as waste products and chemicals. Dissolved substances move in and out of the cell via the cell membrane. There are three main types of cell transport that help with this: diffusion, osmosis, and active transport.

Diffusion

Diffusion is the movement of particles from an area of high concentration to an area of low concentration. It happens in gases and liquids. The difference in concentration is called a **concentration gradient**. Simple sugars, such as glucose, and gases, such as oxygen and carbon dioxide, all move around by diffusion.

Plants also exchange gases via diffusion during photosynthesis and respiration.

The process of diffusion

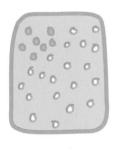  

The orange particles are clumped together when they are first added to the mixture.

As they move around, they collide and start to mix. They move in all directions but the net movement is from high to low concentration.

Diffusion is complete when the orange particles are spread out through the liquid or gas. They continue to move randomly.

Diffusion in the lungs

When you breathe in, oxygen-rich air travels to the lungs.

Oxygen-rich air ends up in tiny air sacs called **alveoli**. This oxygen here is at a high concentration.

Oxygen-rich blood is transported around the body.

Carbon dioxide is exhaled when you breathe out.

Oxygen diffuses out of the alveoli into red blood cells in neighboring blood vessels. Oxygen here is at a lower concentration.

Oxygen-poor blood returns to the lungs from the body.

Carbon dioxide is a waste product of respiration. It diffuses out of red blood cells, where it is in high concentration, into the alveoli, which contain a lower concentration of carbon dioxide.

Osmosis

Osmosis is a special type of diffusion involving water molecules, which occurs when water molecules move from an area where there is a high concentration of water, to an area where there is a low concentration of water. It occurs through a **semipermeable** membrane, which allows only certain substances to pass.

Osmosis also occurs in animals where it helps to maintain the water balance inside cells.

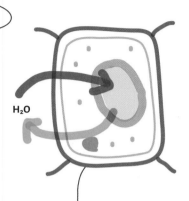

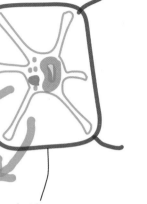

Osmosis in a plant cell

Hypotonic: The water concentration outside the cell is greater than the water concentration inside the cell, so water enters by osmosis. The vacuole swells and puts pressure on the cell wall. This makes the cell firm or turgid.

H_2O

Isotonic: The inside and the outside of the cell have the same concentration of water. There is no net movement of water across the membrane.

Hypertonic: The water concentration inside the cell is greater than the water concentration outside the cell, causing water to leave by osmosis. This causes the vacuole to shrivel and pull away from the cell wall.

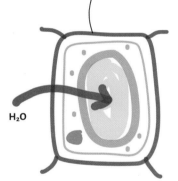

H_2O

Active transport

Active transport is the movement of dissolved substances from an area of low to an area of high concentration, against a concentration gradient. Unlike diffusion and osmosis, this requires energy.

Plant roots contain specialized cells called **root hair cells**. These help the plant to obtain minerals, such as nitrates, from the soil. The minerals move into the root hair cell via active transport.

Active transport also occurs in animals. For example, it is used to transport glucose molecules across the gut wall into the blood.

Active transport in a plant root hair cell

Nitrates in the soil are at a low concentration.

Nitrates in the cell are at a high concentration.

Nitrates enter the cell via active transport.

STEM CELLS AND SPECIALIZATION

Multicellular organisms are made up of lots of different types of specialized cells, such as nerve cells and muscle cells. These specialized cells are formed from unspecialized cells called stem cells. The process of specialization is called **differentiation**.

Stem cells can also divide to produce copies of themselves. Biologists who grow these stem cells in cultures think they could be used to help people who are ill. For example, stem cells could be differentiated into heart muscle cells and used to repair damaged hearts.

In animals, most differentiation occurs early in development, but many plant cells can differentiate throughout life. When a cell differentiates, its DNA stays the same, but key genes are switched on and off. This gives the cell new properties.

Specialized cell development

Red blood cells are specialized cells. They are small and flexible so they can fit through tiny blood vessels. They contain hemoglobin, which binds to oxygen. In mammals, these cells have no nucleus, which frees up room for the hemoglobin. They are slightly flattened, which creates a bigger surface area for oxygen absorption.

Muscle cells are specialized. They are long so they can contract. This takes energy, so they have a lot of mitochondria.

Differentiation is the process by which cells become more specialized. Cells differentiate to acquire different forms and functions. This mainly occurs during development.

The blastocyst develops into an embryo.

Neurons are specialized cells. They carry electrical information around the body. They are long and well insulated. They have branched endings called **dendrites**, which make connections with other cells.

A sperm fertilizes an egg.

A single cell called a **zygote** is created. It contains DNA from both sperm and egg.

The **zona pellucida** is a protective outer layer.

The single cell divides by mitosis to form two cells.

Two cells divide by mitosis to form four cells.

Sperm cells are specialized cells. They have a long tail and a streamlined head to help them swim. They contain lots of mitochondria for energy. The head part contains enzymes, which help break down the outer layers of the egg.

Four cells divide by mitosis to form eight cells, and so on.

The cells form a **blastocyst**. This is a thin-walled, hollow structure containing a cluster of cells.

These cells are embryonic **stem cells**. Stem cells are non-specialized cells. They can differentiate to produce other, more specialized cell types.

Adults have some stem cells, but they're only found in certain places, such as bone marrow. Bone marrow stem cells can differentiate into blood cells, so they are sometimes used to treat leukemia.

The embryo develops into a person. Humans contain billions of cells, made up of hundreds of various specialized, differentiated cell types.

Researchers can make versatile stem cells using adult tissue. Adult skin cells, for example, can be "reprogrammed" to become stem cells, and then used to generate specialized cell types, such as muscle cells. Biologists hope to use these cells therapeutically.

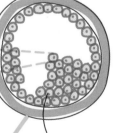

CELL ORGANIZATION

Big, multicellular organisms have many layers of organization, from cells and tissues to organs and organ systems. This organization helps organisms to coordinate many different activities, such as respiration and movement.

Cells to systems

CELLS

Cells are the fundamental building blocks of life. In multicellular organisms, they can't achieve much on their own. So, they work together to carry out important functions.

TISSUES

A tissue is a group of similar cells that work together to carry out a particular function. Sometimes tissues can contain more than one cell type.

A single muscle cell is unable to make a muscle work.

Groups of muscle cells work together to make muscles contract and relax.

A single epithelial cell is unable to form a protective barrier.

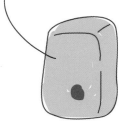

Sheets of epithelial cells offer protection by lining the inside of the intestines, blood vessels, and other structures.

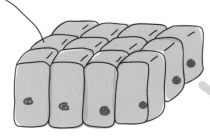

Glandular cells are specialized epithelial cells, but they can't act as glands on their own.

Groups of glandular cells work together to make and release substances, such as enzymes and hormones.

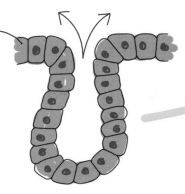

ORGANS

An organ is a group of different tissues that work together to carry out a particular function. For example, the brain is an organ that controls thought, movement, speech, and much more. The heart is an organ that pumps blood around the body. The kidneys filter waste from the blood and make urine. This helps the body get rid of water and excess salts.

The stomach is also an organ. Its job is to digest food. It contains multiple tissue types that work together to achieve this goal. Muscular tissue contracts and relaxes, and helps to mix up the food. Glandular tissue makes digestive juices that help to break down the food. Epithelial tissue covers the inside and the outside of the stomach.

SYSTEMS

Groups of organs are arranged to form systems. The systems work together to carry out particular functions. For example, the respiratory system contains the lungs, the trachea, and various other organs. Its job is to get oxygen into the body, and to remove carbon dioxide from the body.

Different systems depend on one another. For example, the cells of the digestive system rely on the respiratory system for the oxygen that it needs. The cells of the respiratory system rely on the digestive system for the nutrients and energy it needs to function.

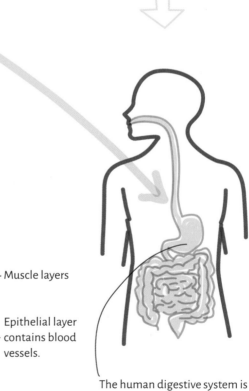

Muscle layers

Epithelial layer contains blood vessels.

Glandular tissue layer

The human digestive system is made up of many different organs, including the stomach.

Organ systems work together to form entire organisms.

LIFE-FORMS

PROKARYOTES

Have cells that lack large organelles and contain free-floating DNA, e.g. bacteria.

EUKARYOTES

Have cells with a membrane-bound nucleus, e.g. fungi, plants, and animals.

UNICELLULAR

Made of a single cell, e.g. amoebas, bacteria.

MULTICELLULAR

Contain lots of cells, e.g. plants, animals.

CELL TYPES

CELLS

ORGANIZATION

CELLS

TISSUES

ORGANS

SYSTEMS

ORGANISMS

STEM CELLS

Can divide to form identical copies.

DIFFERENTIATION

Stem cells can differentiate to produce specialized cells.

SPECIALIZED CELLS

Have particular functions.

MICROSCOPY

MAGNIFICATION

= Image size/real size

LIGHT MICROSCOPES

Low magnification power. Can be used to study live cells.

ELECTRON MICROSCOPES

High magnification power. Cannot be used to study live cells.

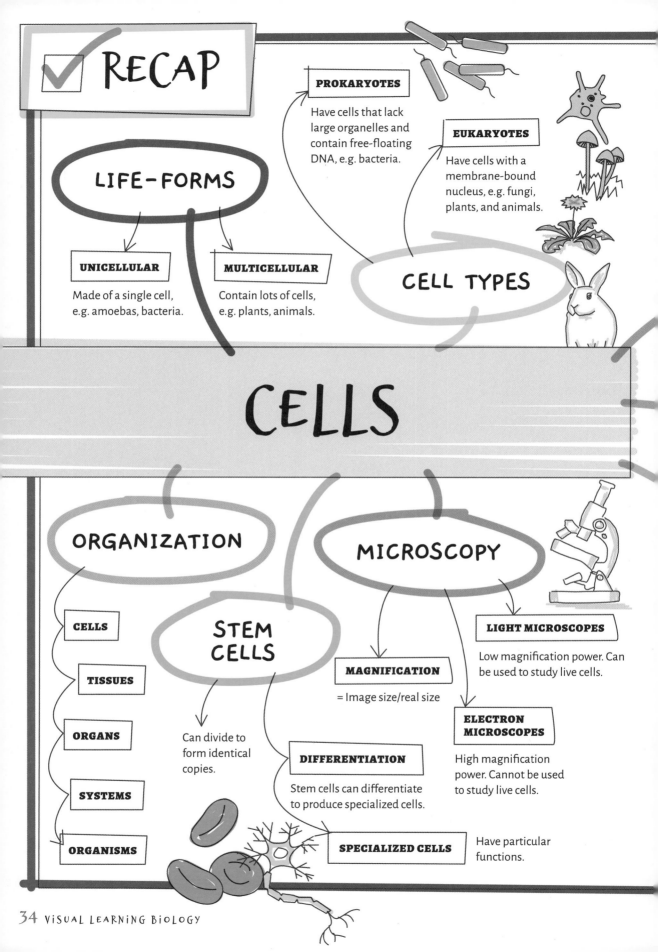

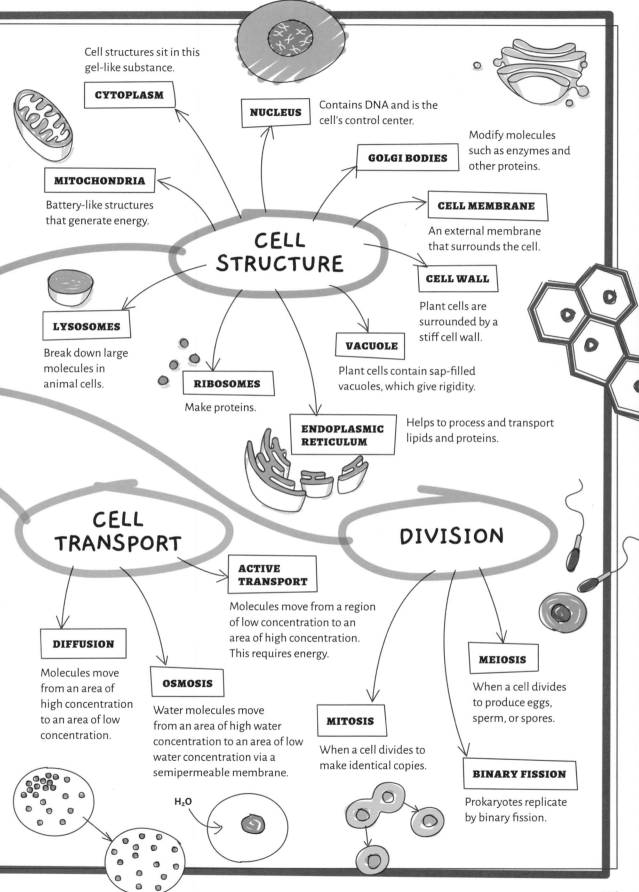

Cell structures sit in this gel-like substance.

CYTOPLASM

NUCLEUS

Contains DNA and is the cell's control center.

Modify molecules such as enzymes and other proteins.

GOLGI BODIES

MITOCHONDRIA

Battery-like structures that generate energy.

CELL MEMBRANE

An external membrane that surrounds the cell.

CELL STRUCTURE

CELL WALL

Plant cells are surrounded by a stiff cell wall.

LYSOSOMES

Break down large molecules in animal cells.

RIBOSOMES

Make proteins.

VACUOLE

Plant cells contain sap-filled vacuoles, which give rigidity.

ENDOPLASMIC RETICULUM

Helps to process and transport lipids and proteins.

CELL TRANSPORT

DIVISION

ACTIVE TRANSPORT

Molecules move from a region of low concentration to an area of high concentration. This requires energy.

DIFFUSION

Molecules move from an area of high concentration to an area of low concentration.

OSMOSIS

Water molecules move from an area of high water concentration to an area of low water concentration via a semipermeable membrane.

H₂O

MEIOSIS

When a cell divides to produce eggs, sperm, or spores.

MITOSIS

When a cell divides to make identical copies.

BINARY FISSION

Prokaryotes replicate by binary fission.

CHAPTER 3

GENETICS

Genetics is the study of DNA and how characteristics, such as height, hair color, and susceptibility to disease are passed from parents to their offspring. All living things inherit genetic information from their parents in the form of DNA.

In this chapter, you'll learn about the structure of DNA, the effects of inherited genes, and what happens when there are mistakes in the genetic sequence. You'll explore the emerging science of gene editing and examine how your DNA interacts with the environment to make you a unique individual.

DNA

Living things carry genetic information in the form of deoxyribonucleic acid, known as DNA. **DNA** is a type of nucleic acid. Nucleic acids are made from the elements carbon, hydrogen, oxygen, nitrogen, and phosphorus.

DNA is also a polymer. **Polymers** are molecules that are made of lots of similar units all bonded together. DNA is made of repeating units called nucleotides. They are arranged in a particular way. This gives DNA its unique structure.

British scientists James Watson and Francis Crick deciphered the structure of DNA in the 1950s. They built a model showing that DNA consists of two strands that are coiled around each other to form a twisted ladder. This is called a **double helix**.

The molecular structure of DNA

Bases: These molecules contain nitrogen. Each base pairs up with another base on the opposite strand of the helix. They always pair up in a particular way. This is known as **complementary base pairing**.

Each **nucleotide** is made of one sugar molecule, one phosphate molecule, and one base.

Phosphate group: This group contains one phosphorus atom and four oxygen atoms.

Sugar: DNA is made from a simple sugar called **deoxyribose**. This is a circular molecule containing five carbon atoms.

There are four types of **base molecule**: thymine (T), adenine (A), cytosine (C), and guanine (G).

Base pairs form the "rungs" of the ladder.

Cytosine

Guanine

Cytosine always pairs with guanine.

Adenine

Thymine

Adenine always pairs with thymine.

Alternating sugar and phosphate molecules form the backbone of the DNA strands. They are the sides of the ladder.

Chromosomes and genes

DNA is a long, thin molecule. Each human cell contains around 2 meters (6.5 feet) of DNA. In humans, DNA can be found inside almost all of the many hundreds of different cell types.

The structure of DNA

Inside the nucleus, DNA is arranged into discrete units called **chromosomes**.

Chromosomes contain **genes**. Genes vary in size. The smallest contain a few hundred bases. The biggest contain more than 2 million bases. Genes are sections of DNA that code for different proteins, such as hemoglobin and testosterone.

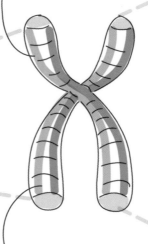

Chromosomes also contain stretches of DNA that don't code for proteins. This is called **noncoding DNA**. Originally, geneticists thought this noncoding DNA had no function, but now they think that most of it is active and has some purpose. Some of it acts like a switch, turning key genes on and off in particular cells. This controls where and when the genes are used, and so influences whether a cell might, for example, become a nerve cell or a muscle cell. Changes in gene activity are referred to as changes in **gene expression**.

1 2 3 4 5
6 7 8 9 10 11 12
13 14 15 16 17 18
19 20 21 22 X Y

Different organisms have different numbers of chromosomes. Humans, for example, have 23 pairs of chromosomes (46 in total). One of each pair of each chromosome comes from the organism's mother, the other comes from its father.

Twenty-two of the human chromosome pairs are **autosomes**. These are any of the chromosomes that are not sex chromosomes. They are labeled from 1 to 22. Autosomes contain lots of different genes, which influence many different features.

The 23rd pair of chromosomes are **sex chromosomes**. They determine whether the offspring is male or female.

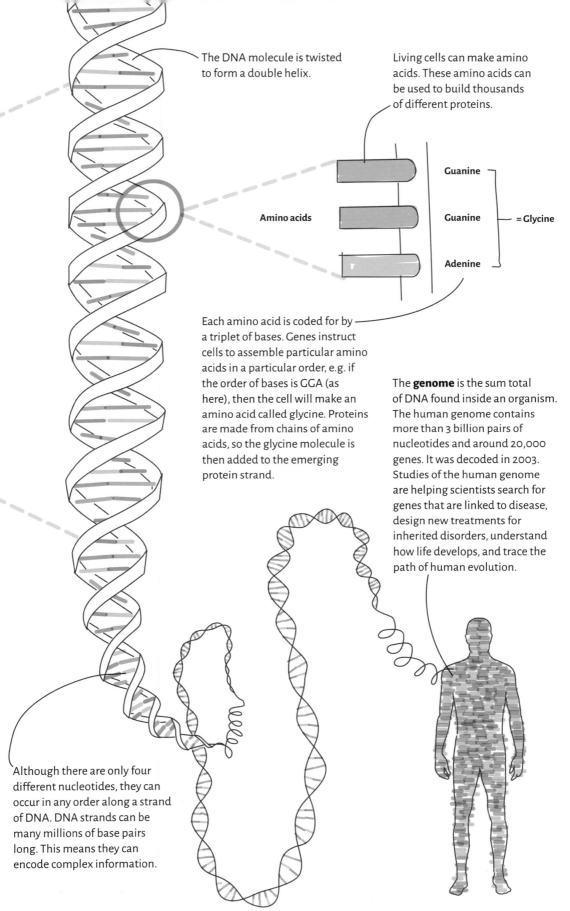

The DNA molecule is twisted to form a double helix.

Living cells can make amino acids. These amino acids can be used to build thousands of different proteins.

Amino acids

Guanine

Guanine   = Glycine

Adenine

Each amino acid is coded for by a triplet of bases. Genes instruct cells to assemble particular amino acids in a particular order, e.g. if the order of bases is GGA (as here), then the cell will make an amino acid called glycine. Proteins are made from chains of amino acids, so the glycine molecule is then added to the emerging protein strand.

The **genome** is the sum total of DNA found inside an organism. The human genome contains more than 3 billion pairs of nucleotides and around 20,000 genes. It was decoded in 2003. Studies of the human genome are helping scientists search for genes that are linked to disease, design new treatments for inherited disorders, understand how life develops, and trace the path of human evolution.

Although there are only four different nucleotides, they can occur in any order along a strand of DNA. DNA strands can be many millions of base pairs long. This means they can encode complex information.

GENETIC INHERITANCE

Individual organisms display characteristics from both parents. This is because they inherit half of their genes from their mother, and half of their genes from their father. Genes influence just about every characteristic you can think of, from your height and eye color to how fast you can run and whether or not you like coffee.

Alleles

Genes come in different versions called **alleles**; for example, there is a gene that controls the color of petals in many flowering plants. One version or allele may result in red flowers. A different allele might produce white flowers. The resulting color of an individual flower is strongly influenced by the alleles that it inherits. Alleles are either dominant or recessive.

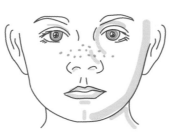

Dominant alleles have an effect even if the individual only has one copy. In humans, for instance, the main gene that causes freckles comes in different alleles. If a person inherits one or two copies of the dominant freckle-causing allele, that individual will have freckles. Dominant alleles are represented with capital letters, as in the diagram on page 41.

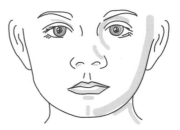

Recessive alleles only have an effect if the individual carries two copies. In humans, there is a recessive allele that leads to the absence of freckles. If someone carries two copies of this allele, that individual will not develop freckles. Recessive alleles are represented with lowercase letters.

If an organism carries two identical alleles of a particular gene, it is said to be **homozygous** for that gene. If an organism carries two different alleles of a particular gene, it is said to be **heterozygous** for that gene.

Predicting genetic outcomes

Gregor Mendel was a nineteenth-century monk who lived in the Czech Republic, then part of Austria. He experimented with pea plants in order to explore the laws that determine genetic inheritance. The results of his research were published in 1866. Today, it forms the basis of modern genetics.

Mendel discovered that "hereditary units" determine plant characteristics. We now know these units are genes. He realized that offspring inherit one unit from each parent, and that the units can be dominant or recessive. This information can be used to predict the outcome of genetic crosses between plant varieties. Today, we use Punnett squares to explain this.

PUNNETT SQUARES

In pea plants, the red allele (R) is dominant, and the white allele (r) is recessive.

This pea plant is homozygous because it contains two alleles that are the same. It has red flowers because the red allele (R) is dominant.

This pea plant is also homozygous, but it has white flowers because it contains two copies of the recessive white allele (r).

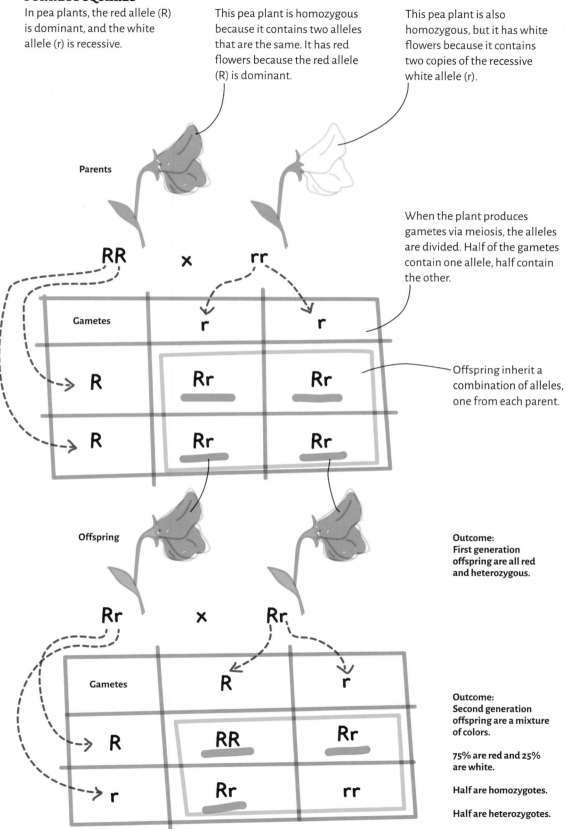

Parents

When the plant produces gametes via meiosis, the alleles are divided. Half of the gametes contain one allele, half contain the other.

RR × rr

Gametes

	r	r
R	Rr	Rr
R	Rr	Rr

Offspring inherit a combination of alleles, one from each parent.

Offspring

Outcome:
First generation offspring are all red and heterozygous.

Rr × Rr

Gametes

	R	r
R	RR	Rr
r	Rr	rr

Outcome:
Second generation offspring are a mixture of colors.

75% are red and 25% are white.

Half are homozygotes.

Half are heterozygotes.

REPRODUCTION

Reproduction is important. It is the mechanism that enables life to persist over time, as organisms pass their genetic information on to their offspring. There are two basic types of reproduction: sexual reproduction and asexual reproduction.

Sexual reproduction

Sexual reproduction occurs when genetic information from two parents is combined. This results in offspring that are genetically different from the parents. Most plants, animals, and fungi reproduce sexually.

Sexual reproduction involves meiosis. This special form of cell division produces gametes that are **haploid**. This means they contain half the normal number of chromosomes. When two

gametes combine to create a zygote, the resulting cell is **diploid**. This means it contains the two complete sets of chromosomes, one from each parent.

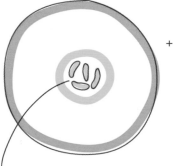
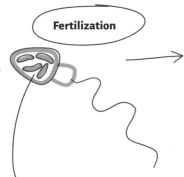
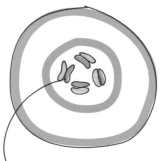

Fertilization

A human egg contains 23 single chromosomes (only four are shown here). It is a haploid cell.

A human sperm contains 23 single chromosomes (only four are shown here). It is a haploid cell.

When the sperm fertilizes the egg, it creates a diploid cell called a **zygote**. It contains 23 pairs of chromosomes, or 46 individual chromosomes. The cell contains a mix of genetic material from the mother and father.

The pros and cons of sexual and asexual reproduction

With sexual reproduction, it takes a lot of time and energy to find a mate and reproduce. The upside is that the offspring inherits a mixture of DNA from both parents. This leads to **genetic**

variation, which means that individuals have slightly different DNA. This is good for the long-term survival of a species; for example, suppose a new infectious disease emerges.

If all the offspring had the same DNA, they might all succumb and die, but if the offspring are genetically varied, some might be able to survive.

The gametes of animals are called sperm and eggs. The gametes of flowering plants are called pollen and eggs. The gametes of fungi are called spores.

Asexual reproduction

Asexual reproduction involves just one parent. The offspring that are produced are genetically identical to the parent. They are known as clones. Prokaryotes, such as bacteria, reproduce by asexual reproduction, as do some plants and animals.

Strawberry plant

Parent plant

Runner

Clone

Strawberry plants reproduce asexually when they put out runners with little plantlets on them.

Daffodil plant

Daffodils grow underground storage organs, which later develop into new plants.

Original plant

Daughter plants form asexually.

The New Mexico whiptail is an all-female species of lizard. It reproduces asexually when an unfertilized egg develops into an adult animal. This is known as **parthenogenesis**. Asexual reproduction does not involve the fusion of gametes. There is no mixing of genetic information, and there is no genetic variation between the parents and their offspring.

Asexual reproduction occurs by mitosis, when a single cell divides to produce an identical copy. This is the same mechanism that is used by animals and plants to make new cells for growth.

New Mexico whiptail lizard

Asexual reproduction only involves one parent. It uses less energy than sexual reproduction because the parent does not have to find a mate. This makes it faster than sexual reproduction. Lots of individuals can be produced quickly, but they are all genetically identical. This is the downside. It means they are less able to deal with the challenges of a changing environment.

SEX DETERMINATION

Just two chromosomes, called the sex chromosomes, determine sex. Each is labeled with a letter. The human sex chromosomes, for example, are referred to as X and Y. Humans have 23 pairs of chromosomes. The 23rd pair is the sex chromosomes. Sex chromosomes are often quite different in size. In humans, the Y chromosome is very small.

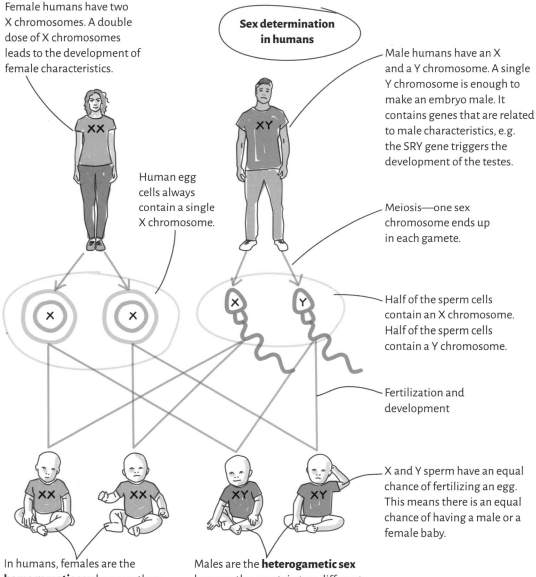

Female humans have two X chromosomes. A double dose of X chromosomes leads to the development of female characteristics.

Sex determination in humans

Male humans have an X and a Y chromosome. A single Y chromosome is enough to make an embryo male. It contains genes that are related to male characteristics, e.g. the SRY gene triggers the development of the testes.

Human egg cells always contain a single X chromosome.

Meiosis—one sex chromosome ends up in each gamete.

Half of the sperm cells contain an X chromosome. Half of the sperm cells contain a Y chromosome.

Fertilization and development

X and Y sperm have an equal chance of fertilizing an egg. This means there is an equal chance of having a male or a female baby.

In humans, females are the **homogametic sex** because they contain two sex chromosomes that are the same.

Males are the **heterogametic sex** because they contain two different sex chromosomes.

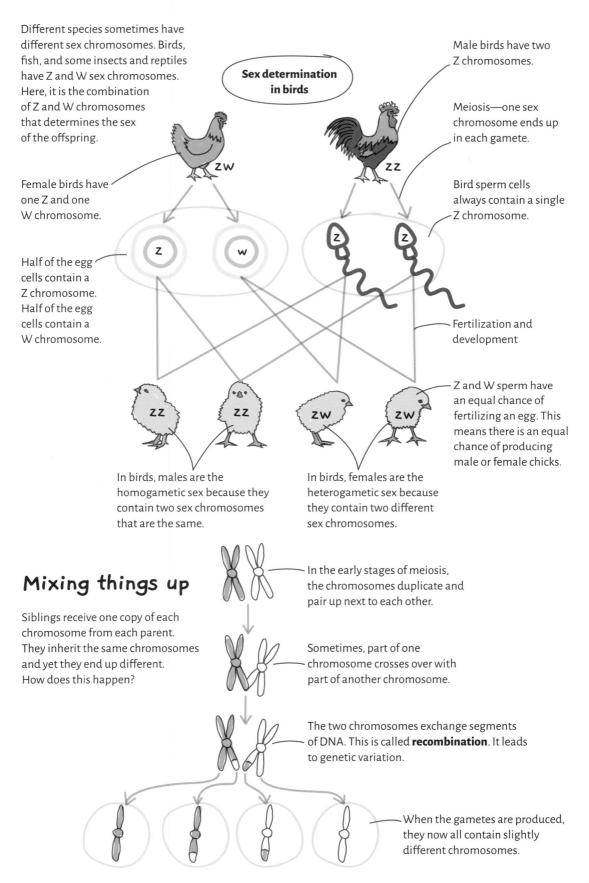

Different species sometimes have different sex chromosomes. Birds, fish, and some insects and reptiles have Z and W sex chromosomes. Here, it is the combination of Z and W chromosomes that determines the sex of the offspring.

Sex determination in birds

Male birds have two Z chromosomes.

Meiosis—one sex chromosome ends up in each gamete.

Female birds have one Z and one W chromosome.

ZW

ZZ

Bird sperm cells always contain a single Z chromosome.

Z   W   Z   Z

Half of the egg cells contain a Z chromosome. Half of the egg cells contain a W chromosome.

Fertilization and development

ZZ   ZZ   ZW   ZW

Z and W sperm have an equal chance of fertilizing an egg. This means there is an equal chance of producing male or female chicks.

In birds, males are the homogametic sex because they contain two sex chromosomes that are the same.

In birds, females are the heterogametic sex because they contain two different sex chromosomes.

Mixing things up

Siblings receive one copy of each chromosome from each parent. They inherit the same chromosomes and yet they end up different. How does this happen?

In the early stages of meiosis, the chromosomes duplicate and pair up next to each other.

Sometimes, part of one chromosome crosses over with part of another chromosome.

The two chromosomes exchange segments of DNA. This is called **recombination**. It leads to genetic variation.

When the gametes are produced, they now all contain slightly different chromosomes.

MUTATIONS

Sometimes the genetic code acquires errors. These errors are called mutations. Some mutations are inherited. They are passed from parent to child. Some mutations occur spontaneously.

Sometimes a mistake is made when the cell is dividing. Sometimes environmental factors, such as pollution or radiation, cause mutations.

Mutations alter the order of bases in the genetic code. This produces genetic variants, where different versions of the same gene exist. Bases code for amino acids, which in turn go on to build proteins. If the mutation causes the amino acid to change, then the protein may be different too. There are different types of mutation: substitution, insertion, and deletion.

Mutations are common. Most of them have little or no effect on the production of proteins. As proteins are big molecules, sometimes small changes do not affect their ability to work properly. Other times, however, the repercussions are large. Some mutations change the shape of the protein and prevent it from working properly. For example, enzymes are complex proteins with very specific shapes. If the shape is altered, then the enzyme may be unable to bind to its target and will not have an effect.

With four bases in the genetic code, there are 64 possible different triplet combinations. However, there are only 20 amino acids. Therefore, sometimes one amino acid can be coded by multiple different triplets.

Mutations

Amino acids have names, such as alanine and threonine. A triplet of bases codes for each amino acid.

Substitution: A single base is substituted for a different base, e.g. a T is swapped for a G. This can make one of the amino acids different.

	Phenylalanine	Threonine	Alanine	
Original code	T T T	A C T	G C A	T
Mutated code	T T G	A C T	G C A	
	Leucine	Threonine	Alanine	

Insertion: An additional base is inserted into the DNA sequence, causing everything downstream to shuffle one place along. This can change lots of amino acids.

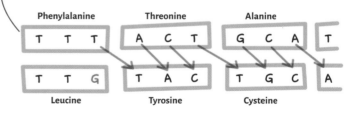

	Phenylalanine	Threonine	Alanine	
	T T T	A C T	G C A	T
	T T G	T A C	T G C	A
	Leucine	Tyrosine	Cysteine	

Deletion: A base is deleted from the sequence, causing everything downstream to shuffle one place along in the opposite direction. This can change lots of amino acids.

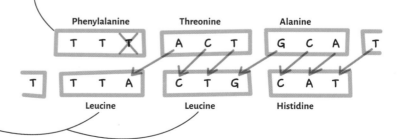

	Phenylalanine	Threonine	Alanine	
	T T T̶	A C T	G C A	T
T	T T A	C T G	C A T	
	Leucine	Leucine	Histidine	

Hereditary disorders

If a mutation is present in a gamete, it can be passed from parent to child. If it has an adverse effect on the person, it can cause a **hereditary disorder**, such as cystic fibrosis or sickle cell disease. Most hereditary disorders are caused by recessive alleles. But some are caused by dominant alleles, for example, Huntington's Disease and Marfan Syndrome.

Cystic fibrosis is caused by a mutation in a key gene. The exact nature of this mutation can vary, but often three nucleotides are deleted. This leads to the production of thick, sticky mucus that clogs the lungs and the digestive system. This causes lung infections and with problems digesting food. Cystic fibrosis is caused by a recessive allele.

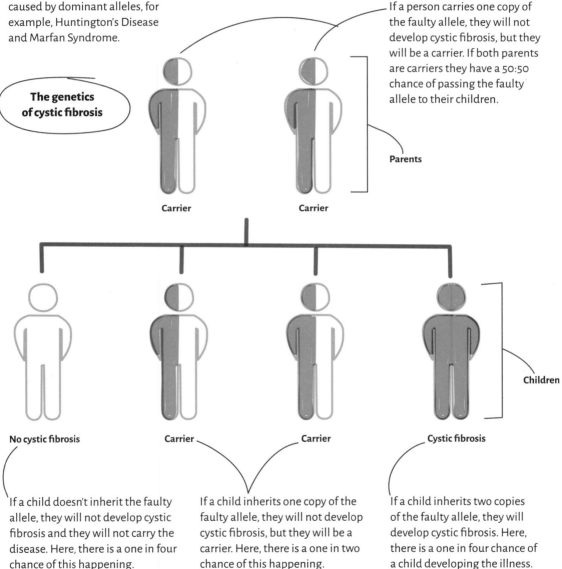

The genetics of cystic fibrosis

If a person carries one copy of the faulty allele, they will not develop cystic fibrosis, but they will be a carrier. If both parents are carriers they have a 50:50 chance of passing the faulty allele to their children.

Parents

Carrier

Carrier

Children

No cystic fibrosis

Carrier

Carrier

Cystic fibrosis

If a child doesn't inherit the faulty allele, they will not develop cystic fibrosis and they will not carry the disease. Here, there is a one in four chance of this happening.

If a child inherits one copy of the faulty allele, they will not develop cystic fibrosis, but they will be a carrier. Here, there is a one in two chance of this happening.

If a child inherits two copies of the faulty allele, they will develop cystic fibrosis. Here, there is a one in four chance of a child developing the illness.

If one or both of the parents already has cystic fibrosis and carries two copies of the faulty allele, then the odds of their offspring inheriting the condition are changed again. If only one parent carries a single faulty gene, the odds of their children developing the disease are also different.

GENE EDITING

At present, hereditary disorders such as cystic fibrosis cannot be cured. The symptoms can be managed to some degree, but the underlying genetic cause remains unchanged. Biologists hope that a new technique called gene editing will change this.

Gene editing with CRISPR-Cas9

Gene editing is a method that can be used to precisely alter an organism's DNA. The most widely used method is called CRISPR-Cas9. It is cheap and easy to use.

CRISPR-Cas9 is like using a pair of molecular scissors. It can be used to snip the DNA strand apart and then insert, remove, or change individual bases of DNA.

While some scientists might find gene editing to have great potential in helping to cure hereditary diseases, ethical concerns arise when it is used to alter human genomes in germline cell therapy. Because of ethical and safety concerns, germline cell and embryo genome editing are currently illegal in many countries.

In the future, CRISPR-Cas9 could be used to edit the DNA inside certain body cells. This is called **somatic cell therapy**. Somatic cells are any of the cells in the body that are not sperm and egg cells. It could be used to alter the DNA inside the lung cells of people with cystic fibrosis. This might ease their symptoms.

CRISPR-Cas9 could also be used to edit the DNA inside sperm and eggs. This is called **germline cell therapy**. It means that any changes will be passed down through subsequent generations. It could be used to alter the gametes of people with the faulty cystic fibrosis allele. This means their offspring would not inherit the disease. It would be a permanent cure for the disorder.

Other uses of gene editing

Gene editing is most widely used in the field of medical research, where it's employed to study gene function, create animal models of disease, and develop new therapies.

It's also being used in food technology to make products, such as gluten-free wheat, naturally spicy tomatoes, and allergy-free foods. In agriculture, gene editing is being used to "beef up" cattle and sheep, and develop strains of animals that are resistant to certain diseases.

Algae are having their DNA edited to help them make energy-efficient biofuels, but the most extreme use of CRISPR-Cas9 is in the field of **de-extinction**. De-extinction is the process of bringing extinct species back to life.

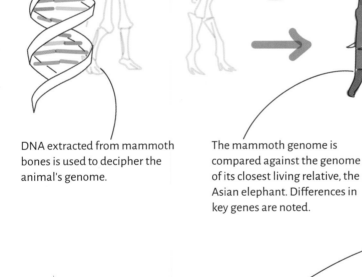

The de-extinction of the woolly mammoth

DNA extracted from mammoth bones is used to decipher the animal's genome.

The mammoth genome is compared against the genome of its closest living relative, the Asian elephant. Differences in key genes are noted.

CRISPR-Cas9 is used to edit these key differences into a living elephant cell.

DNA from the edited cell is used for cloning to produce a living woolly mammoth-like animal. De-extinction technology is still very much in its infancy, so there won't be any woolly mammoth calves soon.

NATURE AND NURTURE

The genomes of non-related people are more than 99.5% identical. The genomes of identical twins are even more similar, and yet we all end up as distinct individuals with our own unique personalities, interests, and ailments. How is this possible if we share so much of our DNA? It all comes down to nature and nurture.

Nature

Living things are influenced by the DNA inside their cells.

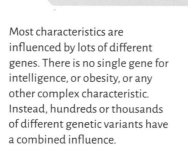

Some characteristics are determined by a single gene, but this is rare. For example, red–green color blindness is caused by a variant in a single gene. A person who has this mutation will struggle to distinguish between reds, greens, and browns.

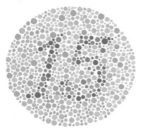

Most characteristics are influenced by lots of different genes. There is no single gene for intelligence, or obesity, or any other complex characteristic. Instead, hundreds or thousands of different genetic variants have a combined influence.

Twin studies

Twin studies are used to tease out the relative importance of nature and nurture.

★ **Identical twins** who are formed when a single fertilized egg splits, share the same DNA.

★ **Non-identical twins** who are formed when separate sperm fertilize separate eggs, share half their DNA.

Twins also tend to grow up in a shared environment. The idea is if a characteristic has a genetic component, then identical twins are more likely to share the characteristic than non-identical twins. There have been thousands of twin studies. They reveal that genes influence just about everything from how fast you can run to whether or not you like coffee. However, the

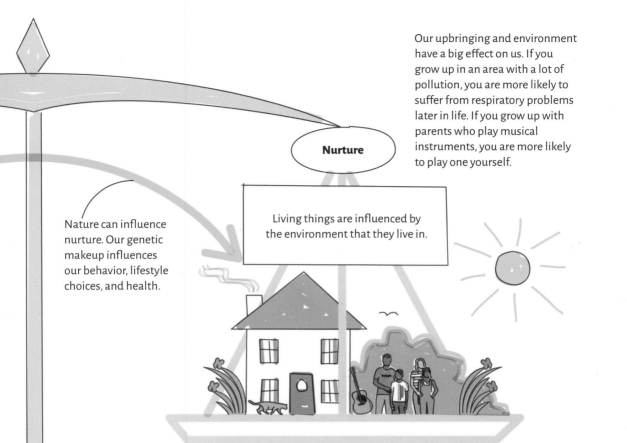

Our upbringing and environment have a big effect on us. If you grow up in an area with a lot of pollution, you are more likely to suffer from respiratory problems later in life. If you grow up with parents who play musical instruments, you are more likely to play one yourself.

Nurture

Nature can influence nurture. Our genetic makeup influences our behavior, lifestyle choices, and health.

Living things are influenced by the environment that they live in.

Nurture can influence nature. The environment "talks" to the body. Environmental factors, such as diet, smoking, and alcohol, can influence the way that genes work. They can turn key genes on and off, but the environmental changes don't change the sequence of the DNA itself. These changes are called **epigenetic**.

studies of twins have shown that genes do not actually determine most of these features. The environment also plays a crucial role. Some features, such as height and eye color, are more influenced by genetics. Others, such as mathematical ability and addiction, are more influenced by the environment in which we are raised. Nature and nurture are both important.

Chance also has a big influence on the way we turn out. Random events can be internal, such as a spontaneous mutation in a certain gene, or external, such as a traffic accident. Events like this can alter our biology and behavior, which in turn can alter patterns of gene activity, which in turn can alter our biology and behavior, and so on.

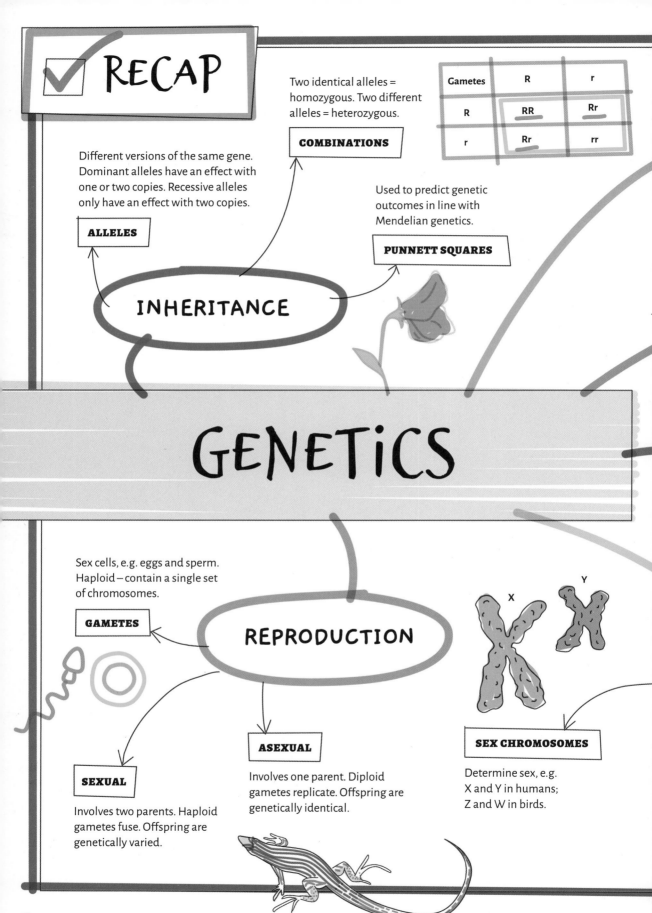

Two identical alleles = homozygous. Two different alleles = heterozygous.

COMBINATIONS

Gametes	R	r
R	RR	Rr
r	Rr	rr

Different versions of the same gene. Dominant alleles have an effect with one or two copies. Recessive alleles only have an effect with two copies.

ALLELES

Used to predict genetic outcomes in line with Mendelian genetics.

PUNNETT SQUARES

INHERITANCE

GENETICS

Sex cells, e.g. eggs and sperm. Haploid – contain a single set of chromosomes.

GAMETES

REPRODUCTION

X   Y

ASEXUAL

SEXUAL

Involves two parents. Haploid gametes fuse. Offspring are genetically varied.

Involves one parent. Diploid gametes replicate. Offspring are genetically identical.

SEX CHROMOSOMES

Determine sex, e.g. X and Y in humans; Z and W in birds.

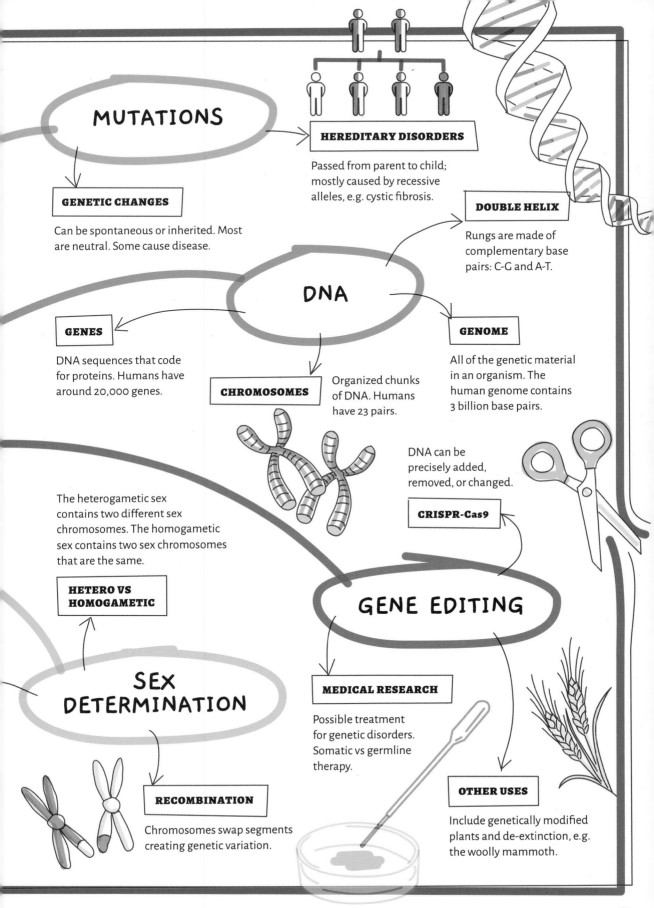

MUTATIONS

HEREDITARY DISORDERS

Passed from parent to child; mostly caused by recessive alleles, e.g. cystic fibrosis.

GENETIC CHANGES

Can be spontaneous or inherited. Most are neutral. Some cause disease.

DOUBLE HELIX

Rungs are made of complementary base pairs: C-G and A-T.

DNA

GENES

DNA sequences that code for proteins. Humans have around 20,000 genes.

CHROMOSOMES

Organized chunks of DNA. Humans have 23 pairs.

GENOME

All of the genetic material in an organism. The human genome contains 3 billion base pairs.

DNA can be precisely added, removed, or changed.

CRISPR-Cas9

The heterogametic sex contains two different sex chromosomes. The homogametic sex contains two sex chromosomes that are the same.

HETERO VS HOMOGAMETIC

GENE EDITING

SEX DETERMINATION

MEDICAL RESEARCH

Possible treatment for genetic disorders. Somatic vs germline therapy.

RECOMBINATION

Chromosomes swap segments creating genetic variation.

OTHER USES

Include genetically modified plants and de-extinction, e.g. the woolly mammoth.

CHAPTER 4

EVOLUTION

The theory of evolution by natural selection is one of the most robust scientific theories that there has ever been. It is underpinned by more than a century of rigorous evidence. It explains how all living things are descended from the same simple life-forms that emerged more than 3 billion years ago, how all of the species on Earth came to be, and how living things continue to change over time. In this chapter, you'll learn more about this fascinating and important theory.

CHARLES DARWIN AND THE VOYAGE OF THE BEAGLE

Two British scientists came up with the theory of evolution independently. They were Charles Darwin and Alfred Russel Wallace. In 1858, they published their observations in separate articles, but both papers were largely overlooked. The following year, Darwin published his book, *On the Origin of Species*, and the idea took off. The theory of evolution became forever linked to his name.

Darwin's finches

Some finches had narrow, pointy beaks that helped them to catch and eat insects.

Darwin's ideas took shape during his travels on HMS *Beagle*. He made detailed observations of the wildlife that he saw as the ship explored the waters around South America.

Darwin studied the finches on the Galápagos Islands. He realized that species varied between islands, and that although some of these species were superficially similar, there were important differences.

Some finches had thick, curved beaks that helped them to crack seeds open.

Galápagos Islands

The Galápagos is an archipelago made up of more than a dozen islands.

South America

Some finches had long, sharp beaks that helped them to tear and eat cactus plants.

Darwin realized that the finches were all closely related and shared a common ancestor. Their differences emerged as they spent time on their different islands and adapted to the conditions there. Along with other experiments and studies of fossils, this led him to propose his "Theory of Evolution via Natural Selection."

EVOLUTION BY NATURAL SELECTION

The theory of evolution, which explains how life changes over time, is one of the most important scientific theories. It states that evolution is underpinned by three vital factors: variation, natural selection, and heritability.

Variation: Members of a species are all broadly similar, but key differences exist. For example, some individuals may be bigger, smaller, more drought-tolerant, or better able to survive the cold. Observable differences like this are called **phenotypes**.

Darwin's theory of evolution by natural selection

Natural selection: The individuals that are best suited to the environment are more likely to survive and reproduce. Darwin called this **survival of the fittest**. The individuals that are less well suited are less likely to survive and reproduce.

Heritability: Individuals pass their successful adaptations to their offspring, who can then pass them on to their offspring and so on. Darwin called this idea **descent with modification**.

Some of these tree beetles have plain gray shells. Others have mottled shells. Features that help organisms to survive are called **adaptations**. The tree beetle's mottled shell is an adaptation.

Mottled shells provide camouflage. Mottled beetles are less likely to get eaten and more likely to reproduce. Plain beetles stand out more. They are more likely to get eaten and less likely to reproduce. The mottled beetles start to become more common and the plain beetles start to become less common.

In time, gray beetles are replaced by mottled beetles. The beetle is evolving.

Today, we know that variation is caused by genetic mutations. Changes to an organism's DNA can lead to changes in phenotype. The resulting adaptations can be useful, harmful, or neutral.

Reactions to Darwin's theory

When it was first published, people made fun of Darwin. Victorian magazines published cartoons showing Darwin's head on a monkey's body.

At the time, the theory of evolution was controversial because:

★ Some people thought Darwin didn't have enough evidence to back up his theory.

★ No one knew exactly "what" was being inherited. Today, we know that genes are inherited when organisms reproduce.

★ It contradicted the then widely held belief that God created all life on Earth.

People misunderstood Darwin. They thought he believed that humans had evolved from monkeys, but Darwin's theory predicted that humans and monkeys share a common ancestor deep in time. This is what many people think today.

Alternatives to Darwin

In the nineteenth century, French biologist Jean-Baptiste Lamarck proposed an alternative to evolution. He thought the way an animal behaves during its lifetime has an effect on its body and that these changes can be inherited. This is called **Lamarckism**. The theory has since been disproved.

Famously, Lamarck suggested that the giraffe evolved its long neck after successive generations reached into the treetops to eat leaves and twigs.

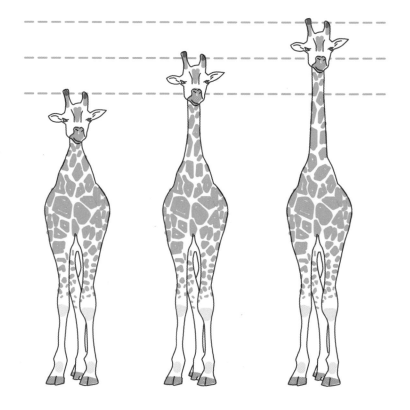

Speciation

Mutations drive variation, and variation generates beneficial adaptations, which help some individuals to survive and reproduce. Over time, this process leads to change and the evolution of new species.

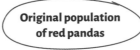

Original population of red pandas

A **species** is a group of similar organisms that can breed with one another and produce fertile offspring. When new species form it is called **speciation**.

New species can form in different ways, but a common way involves separation and isolation.

ISOLATION
Two groups of organisms become physically separated by some sort of barrier, such as a mountain range or a dam. For example, around 250,000 years ago, a single population of red pandas became divided by a river.

VARIATION
Mutations crop up causing variation within the groups. As a result, some red pandas develop different fur color and tail markings.

NATURAL SELECTION
Conditions are different on either side of the river. For example, each region has a slightly different microclimate and geography. In each location, the "fittest" survive and reproduce, and the least "fit" die out. Pandas on one side of the river evolve redder fur and striped tail rings. Pandas on the other side of the river are paler.

HERITABILITY
"Winning" genes get passed down the generations. The genetic makeup of the two populations becomes more varied over time.

SPECIATION
If individuals from the two populations were given the opportunity to mate, they would probably either be uninterested or be unable to produce fertile offspring. Speciation has occurred. Now there are two species of red panda: the Chinese red panda (*Ailurus styani*) and the Himalayan red panda (*Ailurus fulgens*).

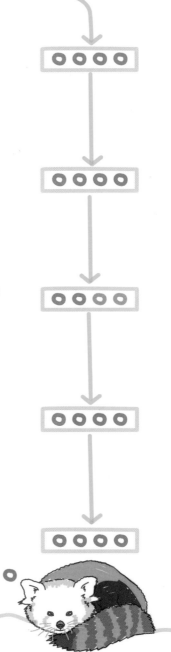

EXTINCTION

New species are born, and old species die out. The **extinction** of a species occurs when there are no more members of that species left. It happens when a species is unable to adapt to environmental change. Scientists estimate that over 99.9% of all species that ever lived have gone extinct.

Extinction can occur for lots of different reasons including:

Asteroids: Dinosaurs went extinct 65 million years ago when an enormous asteroid hit the Earth. This is a very rare occurrence!

Climate change: The Bramble Cay melomys was a rat that lived on a single island in Australia's Great Barrier Reef. It went extinct in 2019.

Disease: In Australia, a marsupial called the Tasmanian devil is at risk because of an infectious cancer that is spread when the animals fight and bite one another.

Invasive species: Non-native species sometimes outcompete native species. Kakapo numbers dropped because invasive species, such as stoats and rats, ate their eggs and chicks.

Human activity: The Yangtze River dolphin went extinct recently after the Chinese rivers it lived in became polluted, overfished, and clogged up with boats.

The kakapo is a ground-dwelling parrot from New Zealand. It is critically endangered, and conservationists are working hard to save it.

Species are at risk of extinction when . . .

★ There are not many individuals left, e.g. there are only around 200 kakapo.

★ The remaining members don't have much genetic variation, e.g. today's kakapo are all descended from a limited number of founding members. Genetic variation is limited because they are closely related.

EVIDENCE FOR EVOLUTION

The theory of evolution by natural selection is now widely accepted. It is the best theory we have to explain the abundance and changing nature of life on Earth. The theory of evolution is now backed up by lots of evidence, collected across time by scientists all over the world.

How do fossils form?

The fossil record

Fossils are the preserved remains of organisms from a long time ago. The **fossil record** is the collective name for all of the fossils that exist. It is important because it provides a series of snapshots of life through time, and it enables scientists to study both extinct species and the process of evolution.

It's rare for an organism to become fossilized, but if the conditions are right, just about any living thing can be preserved in stone. Scientists have found fossils of animals, plants, fungi, and even bacteria. Organisms can also be preserved in peat bogs, tar pits, amber, and in ice.

The fossil record is frustratingly patchy and incomplete. Despite this, scientists have found many amazing fossils. They show how life changes over time.

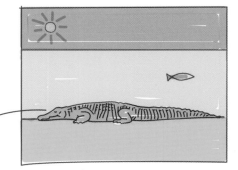

A crocodile dies and sinks to the bottom of the river.

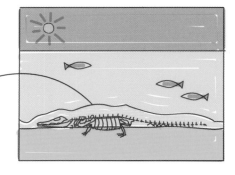

The dead animal is quickly buried by small particles of rock called **sediment**. The soft parts rot away, and the hard parts, such as bone and teeth, are left behind.

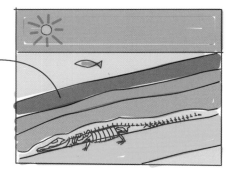

More layers of sediment pile up on top. This presses down on the skeleton. Minerals seep into the bone and replace the biological molecules that the crocodile was made from. Over millions of years, the bone is turned to stone.

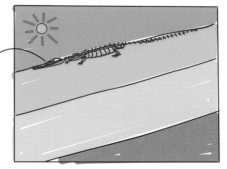

Geological processes cause the seabed to rise up and become exposed. The layers of rock get worn away by wind and rain. This is called **erosion**. The fossil is finally revealed.

Increasing complexity of life

The theory of evolution predicts that all life on Earth is descended from simple organisms that emerged billions of years ago, and that, as evolution progressed, life became more complex. This idea is supported by the fossil record. It predicts that species go extinct. The fossil record contains many extinct species, for example *Tyrannosaurus rex*. It also predicts the existence of **transitional fossils**; intermediate life-forms with features of distant ancestors and more recent descendants.

More complicated life-forms are found in younger rocks.

The simplest life-forms are found in the oldest rocks.

Archaeopteryx is one of the world's most famous fossils. This raven-sized animal lived around 150 million years ago.

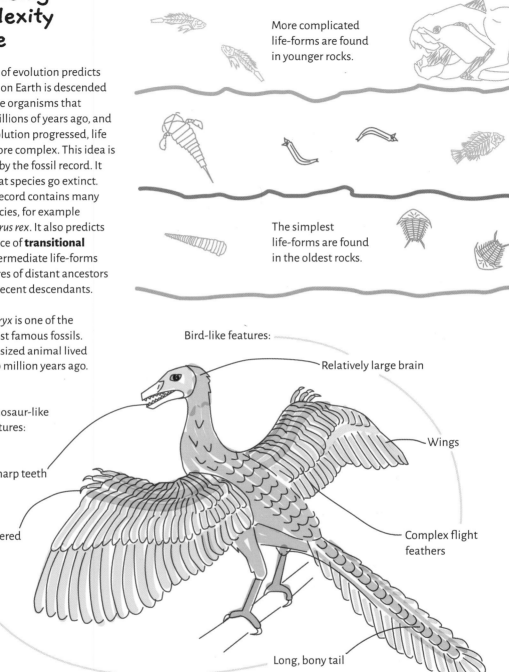

Bird-like features:

Relatively large brain

Wings

Complex flight feathers

Dinosaur-like features:

Jaw with sharp teeth

Three-fingered claw

Long, bony tail

When Darwin published *On the Origin of Species* in 1859, he was troubled by the lack of transitional fossils. People found it hard to imagine, for example, how dinosaurs, which are reptiles, could transform into something as different as birds.

Two years later, the first fossils of *Archaeopteryx* were found. *Archaeopteryx* is important because it provides a "missing link" between dinosaurs and birds, and it lends support to Darwin's theory. It's now widely accepted that birds are descended from a group of two-legged dinosaurs called theropods, and since then many other transitional fossils have been found.

Evolution in action

When Darwin thought about evolution, he imagined that changes happen slowly over thousands and millions of years, but sometimes we can see evolution happening right in front of our eyes. It provides us with more evidence for the theory of evolution.

THE PEPPERED MOTH
The peppered moth is an evolutionary icon. It clearly shows how organisms adapt and evolve in response to a changing environment. It's also known as "Darwin's moth" because it supports the theory of evolution.

The peppered moth flies at night and rests during the day.

Peppered moths are cream with black spots.

Soot and smoke from the factory chimneys settle on the tree trunks where the peppered moths rest during the day.

A random mutation alters a gene involved in pigmentation. Moths with the mutation develop black wings.

The peppered moth evolves. The rare black form becomes more common because it camouflages better. The cream form becomes less common because birds can spot it more easily and eat it.

The Clean Air Act is passed. There is less pollution and the tree trunks return to their normal color.

The peppered moth evolves. The black form becomes less common because birds can spot it more easily and eat it. The cream form becomes more common because it camouflages better.

RESISTANT BACTERIA

It's relatively easy to see evolution occurring in small organisms that reproduce quickly. Bacteria are a good example of this. The emergence of antibiotic resistance in bacteria is another example of evolution by natural selection.

Antibiotics are prescribed by doctors and widely employed in agriculture, where they are used to prevent diseases in farm animals and crops.

Antibiotics are only effective against bacteria, but they're often used when they are not needed. This increases the risk of resistant strains of bacteria developing, which cannot be treated by antibiotics. Methicillin-resistant Staphylococcus aureus (MRSA) is one of these. It is dangerous because it is resistant to most antibiotics.

To reduce the rise of antibiotic-resistant bacteria:

★ Doctors should only prescribe antibiotics for serious bacterial infections—not for viruses.

★ Patients should take the full course of antibiotics that they are prescribed. This makes sure that all the bacteria are killed, so they can't mutate and form resistant strains.

★ Antibiotics should be used less in agriculture.

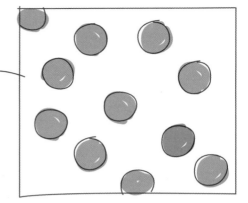

Normal bacteria—one of the bacterial cells (in red) contains a random mutation.

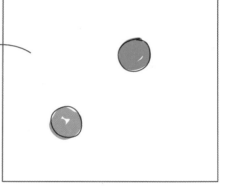

The bacteria are treated with an antibiotic. It kills most of the normal bacteria, but not the one with the mutation. This bacterium is resistant to the antibiotic.

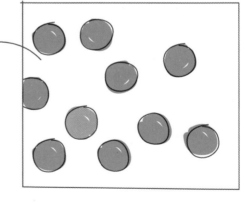

Bacteria without the mutation either die or are unable to reproduce. The resistant bacteria multiply and become more common.

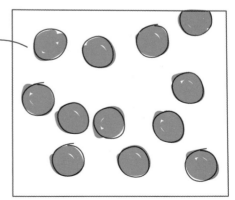

A new antibiotic-resistant strain of bacteria has emerged. It spreads because people have no immunity to it and because antibiotics can't kill it.

 Normal bacterium

Resistant bacterium

Shared anatomical features

If all life on Earth is descended from the same common ancestor, as evolutionary theory predicts, then related organisms should share certain anatomical characteristics. When different species share similar anatomical features that they inherited, they are called **homologous structures**.

Humans, birds, and whales all share a very distant common ancestor, so although they look very different on the outside, on the inside there are similarities.

These bones are all organized in a similar way. They are homologous structures.

Bird wing

Human arm

Whale flipper

> **Homologous structures**

Vertebrates are animals with backbones. They are all descended from the same common ancestor, and as a result, vertebrate embryos all look very similar. All vertebrate embryos have a tail. In some animals, such as fish, the tail develops, but in others, such as humans, it doesn't. In humans, the tail is also a vestigial structure. **Vestigial structures** are inherited features that serve little or no present purpose. In the past, however, they were useful.

> **Vertebrate embryos with tails**

Human

Fish

Chicken

Pig

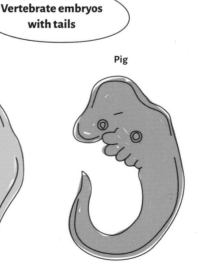

Shared genetic features

All living things use the same genetic code, which is passed down through the generations from parent to offspring. The theory of evolution predicts that all life on Earth is descended from the same common ancestor. If this is the case, then living things should share genetic features too.

Geneticists have compared the genomes of many different species, and found that organisms share many of their genes. For example, humans share thousands of their genes with other organisms, including bacteria and plants.

Closely related species share more of their genes than more distantly related species. For example, we humans share more genes with apes, such as gorillas and chimpanzees, than we do with other species. This suggests that apes are our closest relatives. This finding is backed up by the fossil record, which also demonstrates a close link between humans and apes.

Researchers draw **evolutionary trees** to show how life evolves and how different species are related to one another.

Evolutionary trees, such as the one below, are simplistic. Sometimes species interbreed, so there are also connections between branches. In reality, the tree of life is less like a tree and more like a bramblebush!

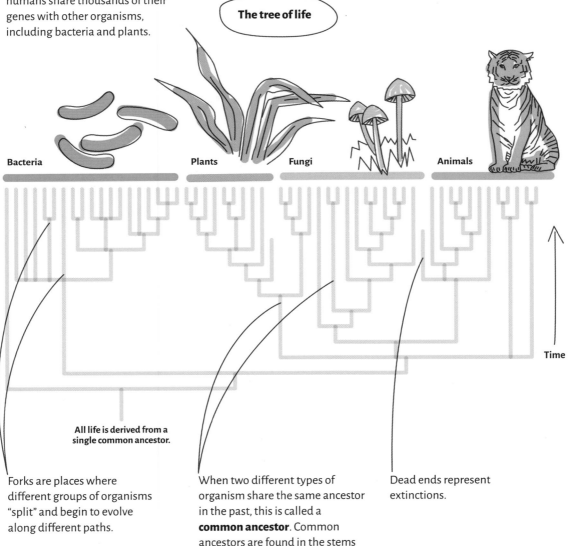

The tree of life

Bacteria

Plants

Fungi

Animals

Time

All life is derived from a single common ancestor.

Forks are places where different groups of organisms "split" and begin to evolve along different paths.

When two different types of organism share the same ancestor in the past, this is called a **common ancestor**. Common ancestors are found in the stems of the evolutionary tree.

Dead ends represent extinctions.

HUMAN EVOLUTION

The story of human evolution is complicated, long, and being refined all the time as new evidence comes to light. Our species is *Homo sapiens*. We call ourselves "modern humans" because there were lots of related human and human-like species that came before us.

Timeline of how humans evolved

Seven million years ago, humans and chimpanzees shared a common ancestor. When this group split, some of the descendants developed into modern apes. The remainder went on to become human. We call them, and all of their descendants, **hominins**.

Australopithecines were the first hominins. After them, many more followed. Our group or genus, known as *Homo*, emerged between 2.4 and 1.5 million years ago. There were at least nine different species of *Homo* including *Homo habilis*, *Homo erectus*, and *Homo neanderthalensis* (Neanderthals).

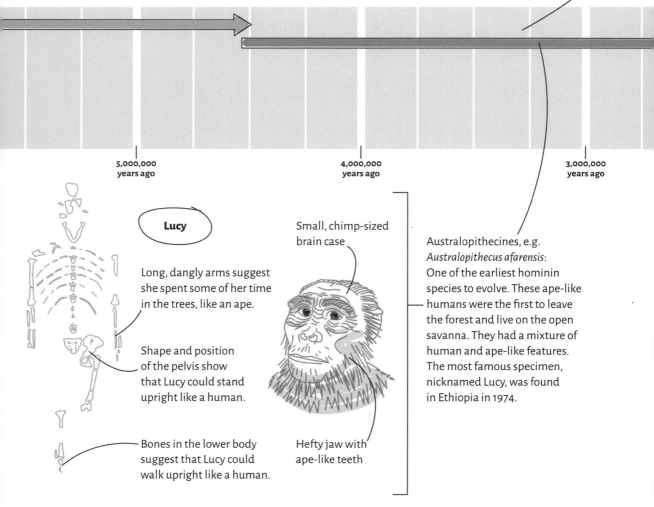

5,000,000 years ago

4,000,000 years ago

3,000,000 years ago

Lucy

Long, dangly arms suggest she spent some of her time in the trees, like an ape.

Shape and position of the pelvis show that Lucy could stand upright like a human.

Bones in the lower body suggest that Lucy could walk upright like a human.

Small, chimp-sized brain case

Hefty jaw with ape-like teeth

Australopithecines, e.g. *Australopithecus afarensis*: One of the earliest hominin species to evolve. These ape-like humans were the first to leave the forest and live on the open savanna. They had a mixture of human and ape-like features. The most famous specimen, nicknamed Lucy, was found in Ethiopia in 1974.

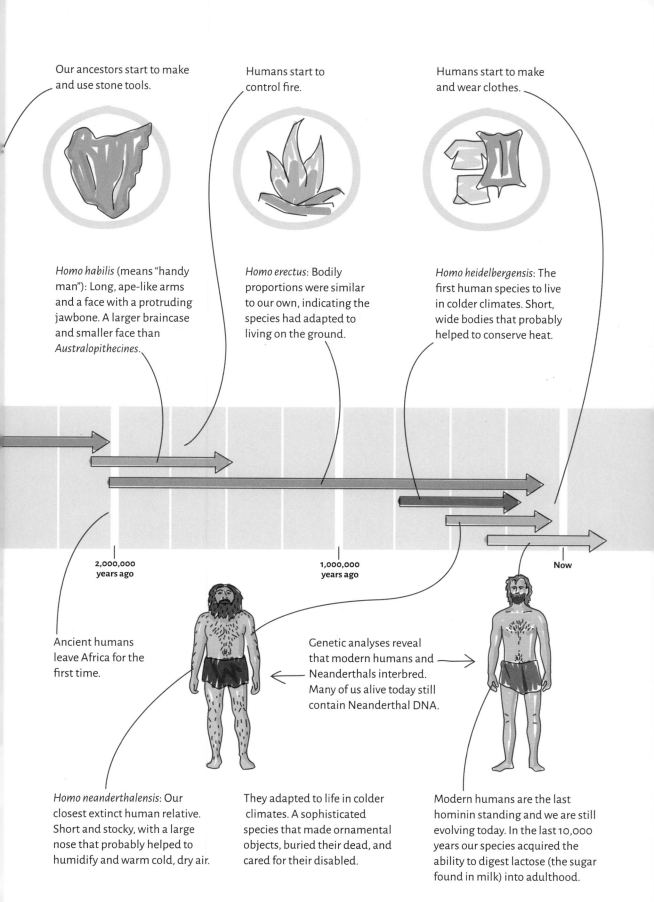

Our ancestors start to make and use stone tools.

Humans start to control fire.

Humans start to make and wear clothes.

Homo habilis (means "handy man"): Long, ape-like arms and a face with a protruding jawbone. A larger braincase and smaller face than *Australopithecines.*

Homo erectus: Bodily proportions were similar to our own, indicating the species had adapted to living on the ground.

Homo heidelbergensis: The first human species to live in colder climates. Short, wide bodies that probably helped to conserve heat.

2,000,000 years ago

1,000,000 years ago

Now

Ancient humans leave Africa for the first time.

Genetic analyses reveal that modern humans and Neanderthals interbred. Many of us alive today still contain Neanderthal DNA.

Homo neanderthalensis: Our closest extinct human relative. Short and stocky, with a large nose that probably helped to humidify and warm cold, dry air.

They adapted to life in colder climates. A sophisticated species that made ornamental objects, buried their dead, and cared for their disabled.

Modern humans are the last hominin standing and we are still evolving today. In the last 10,000 years our species acquired the ability to digest lactose (the sugar found in milk) into adulthood.

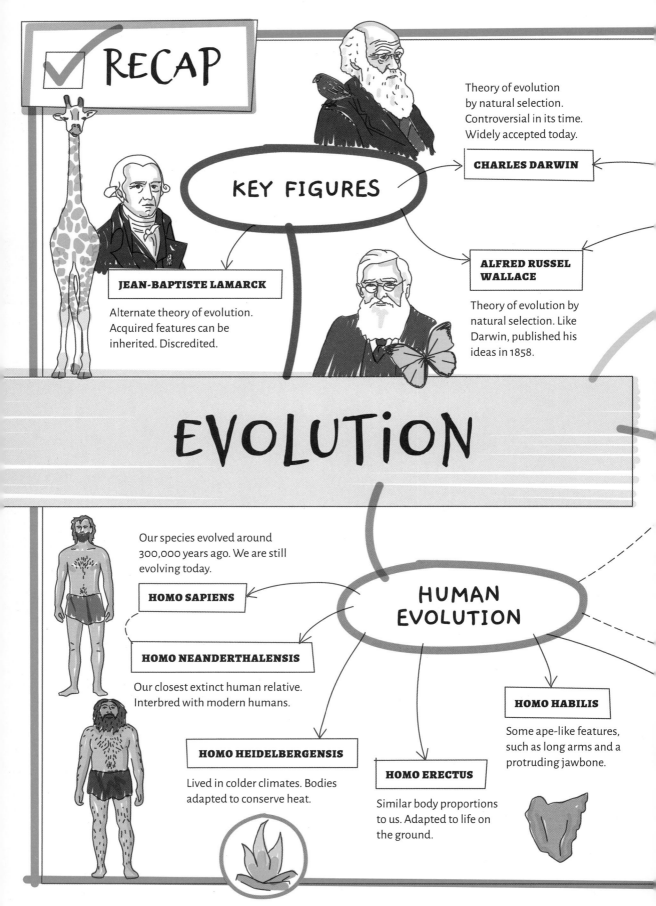

✓ RECAP

KEY FIGURES

CHARLES DARWIN

Theory of evolution by natural selection. Controversial in its time. Widely accepted today.

JEAN-BAPTISTE LAMARCK

Alternate theory of evolution. Acquired features can be inherited. Discredited.

ALFRED RUSSEL WALLACE

Theory of evolution by natural selection. Like Darwin, published his ideas in 1858.

EVOLUTION

HUMAN EVOLUTION

HOMO SAPIENS

Our species evolved around 300,000 years ago. We are still evolving today.

HOMO NEANDERTHALENSIS

Our closest extinct human relative. Interbred with modern humans.

HOMO HEIDELBERGENSIS

Lived in colder climates. Bodies adapted to conserve heat.

HOMO ERECTUS

Similar body proportions to us. Adapted to life on the ground.

HOMO HABILIS

Some ape-like features, such as long arms and a protruding jawbone.

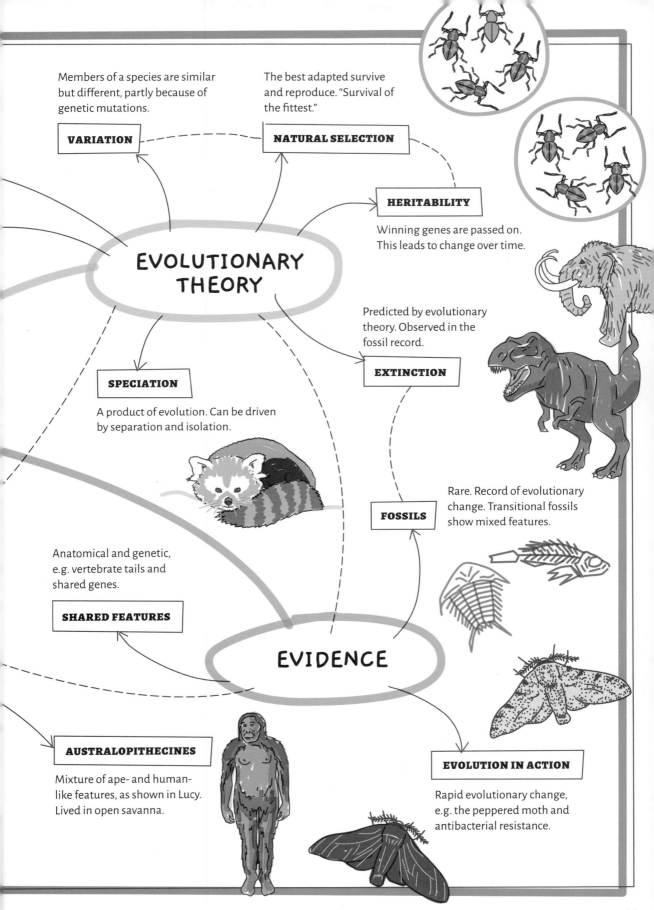

Members of a species are similar but different, partly because of genetic mutations.

VARIATION

The best adapted survive and reproduce. "Survival of the fittest."

NATURAL SELECTION

EVOLUTIONARY THEORY

HERITABILITY

Winning genes are passed on. This leads to change over time.

Predicted by evolutionary theory. Observed in the fossil record.

EXTINCTION

SPECIATION

A product of evolution. Can be driven by separation and isolation.

Rare. Record of evolutionary change. Transitional fossils show mixed features.

FOSSILS

Anatomical and genetic, e.g. vertebrate tails and shared genes.

SHARED FEATURES

EVIDENCE

AUSTRALOPITHECINES

Mixture of ape- and human-like features, as shown in Lucy. Lived in open savanna.

EVOLUTION IN ACTION

Rapid evolutionary change, e.g. the peppered moth and antibacterial resistance.

CHAPTER 5

ORGANIZING LIFE

Scientists estimate there are around 9 million different species living on our planet. These range from simple, single-celled bacteria to complex, multicellular organisms, such as plants and animals— and most have yet to be described. Scientists like to organize life into different groups based on the various features that organisms have. This is called classification. In this chapter, you'll explore the diversity of life on Earth and learn how it is classified.

THE IMPORTANCE OF CLASSIFICATION

The scientific name for classification is **taxonomy**. The groups that living things are assigned to are called **taxonomic groups**, for example, a species is one type of taxonomic group.

Biologists classify living things for many different reasons.

★ It helps to make sense of the enormous variety of living things that exist.

★ It helps to figure out how different living things are related, and how evolution occurs.

★ It helps biologists to understand newly discovered organisms. For example, if a new species of bacteria is found, scientists can compare it to other known bacteria and then gauge how infectious or not it might be.

★ It provides biologists with a shared language. Scientists all over the world refer to particular species by particular names. This helps to avoid confusion.

Bacteria: True bacteria are simple, microscopic organisms made of single cells.

LIFE

Eukaryotes: A complex, broad range of organisms, including animals, fungi, plants, and protists.

Archaea: A type of primitive bacteria often found in extreme environments.

Three-domain system

In the past, life was classified in different ways, but today, living things are grouped into one of three domains: bacteria, archaea, and eukaryotes. After that, they are subdivided into progressively smaller groups, such as kingdoms, families, and species.

Carl Woese, an American scientist, proposed the three-domain system in the 1970s.

CLASSIFICATION

Life is classified according to a system that was proposed in the 1700s by a Swedish scientist named Carl Linnaeus. It is called the **Linnaean system**, and it's made up of broad categories that are subdivided into smaller, increasingly refined groups. Kingdoms are subdivided into phyla, which are further subdivided into classes, then orders, families, genera, and species.

Scientists group things together on the basis of shared features. Organisms that share more features are more closely related to one another than organisms that share fewer features.

In the past, scientists classified organisms by comparing obvious features, such as structure and function. Today, they use more sophisticated methods, including genetic and chemical analyses. As a result, they are continually redefining how life is classified.

Some organisms also have a common name. *Canis lupus* is the gray wolf. It is a large canine found in North America and Eurasia. Common names can vary around the world.

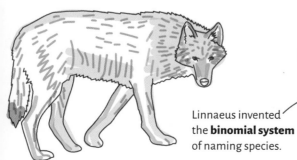

Linnaeus invented the **binomial system** of naming species.

Increase in numbers. Decrease in similarity.

KINGDOM: ANIMALIA
Members eat organic material, breathe oxygen, and can move.

PHYLUM: CHORDATA
Members have a spine made of cartilage or bone.

CLASS: MAMMALIA
Members are warm-blooded, have hair or fur, mammary glands, and four-chambered hearts.

ORDER: CARNIVORA
Members eat meat. They have claws and teeth adapted for catching and eating their prey.

FAMILY: CANIDAE
Members are all dog-like. They include wolves, dogs, jackals, and related animals.

GENUS: CANIS
Members vary greatly in size. They tend to have well-developed skulls and teeth and long legs.

SPECIES: *CANIS LUPUS*
Each organism has a two-part Latin name. The first part is the genus. The second part is the species. Wolves are *Canis lupus*.

KINGDOM

There are five different kingdoms. They are based on development and nutrition, e.g. animals and plants are separate kingdoms because animals have to eat food, and plants make their own food.

PHYLUM

Phyla (plural of phylum) are based on features that define the groups within the kingdom, e.g. flowering plants and cone-bearing plants are in separate phyla.

CLASS

Classes are based on features that define the groups within the phylum, e.g. bivalves (clams, mussels) and gastropods (snails, slugs) are in separate classes within the phylum Mollusca.

ORDER

Orders are based on features that define the groups within the class, e.g. scorpions and spiders are in separate orders within the class Arachnida.

FAMILY

Families are based on key features that define the groups within the order, e.g. lemurs and great apes are in different families within the order of Primates.

GENUS

Genera (plural of genus) are based on features that define the groups within the family, e.g. roses and cherries are in different genera within the family Rosaceae.

SPECIES

Members of the same species can interbreed and produce fertile offspring.

Decrease in numbers. Increase in similarity.

FIVE KINGDOMS

Animals
Multicellular creatures, e.g. tigers

Plants
Multicellular, mainly photosynthetic organisms

Fungi
Organisms, such as mushrooms, molds, and yeast

Protists
Eukaryotic organisms that are not animals, plants, or fungi, e.g. amoeba and *Plasmodium*.

Prokaryotes
Single-celled organisms lacking large, membrane-bound organelles, e.g. bacteria and blue-green algae.

Members of individual species can vary a great deal. So, brown, egg-laying chickens look very different to small, fluffy, ornamental bantams.

PROKARYOTES

Prokaryotes are simple, unicellular organisms with nucleic acids that float freely in the cytoplasm. Prokaryotes may be tiny, but they are incredibly important. If humans disappeared overnight, life would continue. If all the prokaryotes disappeared, life would stop. Prokaryotes play many vital roles, such as recycling nutrients, converting molecules into useful biological forms, and unlocking supplies of carbon, nitrogen, and other elements.

Common bacterial shapes

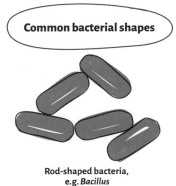

Rod-shaped bacteria, e.g. *Bacillus*

Spherical bacteria, e.g. *Coccus*

Spiral-shaped bacteria, e.g. *Spirillum*

Bacteria

Bacteria are the most diverse and widespread of all the prokaryotes. They are found almost everywhere on Earth. Some bacteria even thrive in extreme environments, such as acidic hot springs and radioactive waste. Bacteria have been found deep down in the Earth's crust and high up in the Earth's atmosphere.

Bacteria come in many different shapes, such as rods, spheres, and spirals. Some bacteria, such as *E. coli* and *Staphylococcus*, are well-known, but most bacteria have yet to be characterized.

Bacteria are referred to as **true bacteria** to distinguish them from archaea, which can look superficially similar. Bacteria were among the first life-forms to evolve.

Most animal life depends on bacteria for survival. This is because bacteria contain the genes and enzymes needed to make vitamin B_{12}. They provide it to animals via the food chain.

Some bacteria are harmful and cause disease, but many bacteria are beneficial. The bacteria that live in our guts, for example, help to digest food and act as a barrier against intestinal infection.

GROWING BACTERIA
Bacteria can be grown in the lab. This is useful for scientists who want to study them and develop new antibiotics.

A petri dish is filled with a mixture of nutrients. This is the **culture medium**. It can be a solid jelly, such as agar, or a liquid broth.

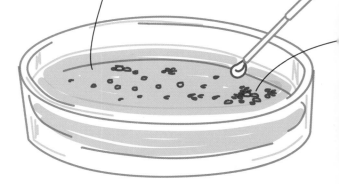

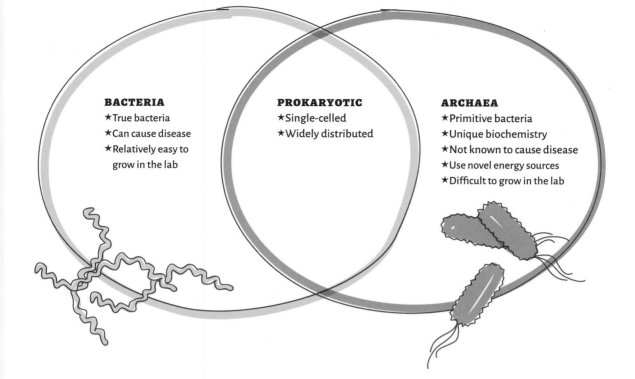

BACTERIA
★True bacteria
★Can cause disease
★Relatively easy to
  grow in the lab

PROKARYOTIC
★Single-celled
★Widely distributed

ARCHAEA
★Primitive bacteria
★Unique biochemistry
★Not known to cause disease
★Use novel energy sources
★Difficult to grow in the lab

Bacteria are transferred to the culture medium using a pipette or a wire loop.

The petri dish is transferred to a warm, moist incubator, where the bacteria multiply. If the bacteria are growing on agar, small visible colonies will form on the surface. If the bacteria are grown in a liquid, the multiplying bacteria form a cloudy suspension.

Archaea

Archaea have been found in extreme environments, such as volcanic springs and deep-sea hydrothermal vents. Scientists originally classed them as **extremophiles** (organisms that thrive in extreme environments), but now we realize that they are much more widespread.

Archaea tend to be even smaller than bacteria. They are referred to as **primitive bacteria**, but are actually more closely related to eukaryotes than to true bacteria.

Archaea are subdivided into many different phyla, but it's hard to classify them because they are very difficult to grow and study in the lab.

Although bacteria and archaea look alike, there are important biochemical and molecular differences between them. For example, the cell membranes of archaea are made of a different kind of lipid.

Archaea can also use novel energy sources. These range from organic compounds, such as sugars, to inorganic substances, such as metal ions, hydrogen gas, and the sun's light rays.

EUKARYOTES

Eukaryotes are the most diverse domain of life. Eukaryotes include all of the large, multicellular life-forms that we are familiar with, such as animals and plants, and this domain also contains many simple, single-celled organisms, such as phytoplankton.

Eukaryotes are subdivided into four kingdoms: animals, plants, fungi, and protists. Together, eukaryotes make up more than 85% of the world's biomass.

Protists

Most protists are simple, single-celled eukaryotes. They include organisms, such as amoebas, diatoms, and slime molds. Protists come in many shapes and sizes. They can be regularly or irregularly shaped. Some are covered in tiny hair-like structures, called cilia, which propel them along. Others have whiplike tails called flagella. Protists are more diverse in shape and function than animals, fungi, and plants.

Protists obtain their nutrients in different ways. Some are **autotrophic**; they make their own food using simple substances found in the environment. Some use light as an energy source (photosynthesis), while others use inorganic chemicals (chemosynthesis). Some are **heterotrophic**; they cannot make their own food and instead obtain energy by taking in organic substances. Some do a bit of both. They are called **mixotrophs**.

Some protists are **parasites**. This means they live on or in other organisms and cause damage. Very often, the parasite is transferred to the host by a third party, which doesn't get any disease that is caused by the parasite. This organism is called a **vector**; for example, mosquitoes are the vector for malaria. When they bite, they transmit a disease-causing parasite called *Plasmodium*.

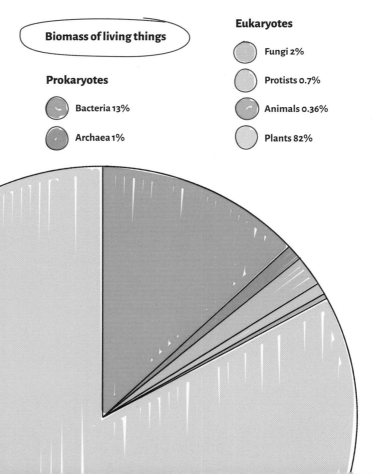

Biomass of living things

Prokaryotes

Bacteria 13%

Archaea 1%

Eukaryotes

Fungi 2%

Protists 0.7%

Animals 0.36%

Plants 82%

SPOTLIGHT ON PROTISTS: SLIME MOLDS

Slime molds are protists. There are more than 900 different species of slime mold. They are often found growing on forest floors, on rotting logs, and in the mulch that collects in rain gutters. They play an important ecological role because they help to break down decaying plant matter, and they feed on bacteria, yeast, and fungi.

When food is plentiful, slime mold often live freely as single cells.

The spores are released. They hatch and turn into single-celled slime molds.

The amoeba-like cells reproduce and produce zygotes.

Slime molds are remarkable. When a slime mold mass is minced into pieces, the separated pieces can find one another and reform. Although they have no brains, slime molds can learn. For example, the yellow slime mold *Physarum polycephalum* can solve mazes. It also comes in over 700 different sexes, which are determined by the variants of key genes that are inherited by the slime mold.

When food is sparse, the slimy mass produces rigid fruiting bodies that contain spores.

The cells join together to form a single slime-like mass. It can be anything from a few centimeters to a few meters in size. The mass contains lots of nuclei. It can detect and engulf microorganisms.

Fungi

Fungi are one of the four different eukaryotic kingdoms. Yeasts, mushrooms, and toadstools are all fungi. They can't make their own food like plants do, and they can't consume food like heterotrophic animals. Instead, fungi are **heterotrophic decomposers**. They obtain nutrients by absorbing dissolved substances.

The cell walls of fungi contain a strong, nitrogen-containing polysaccharide called chitin.

Fungal vacuoles help store small molecules and regulate water concentration inside the cell.

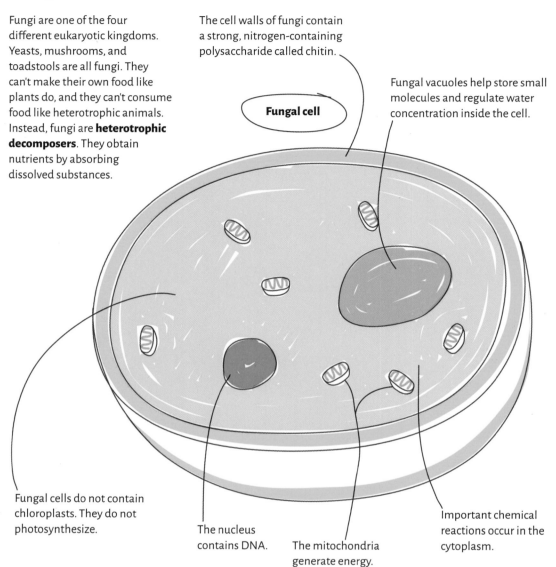

Fungal cell

Fungal cells do not contain chloroplasts. They do not photosynthesize.

The nucleus contains DNA.

The mitochondria generate energy.

Important chemical reactions occur in the cytoplasm.

Fungi are abundant worldwide. Some are single-celled, like yeast. Others, like the mushrooms we are familiar with, are multicellular.

Some fungi are parasitic, e.g. the fungus that causes mildew in sugar beet plants. Others form mutually beneficial **symbiotic** relationships with other living things, such as plants and animals.

Fungi are important ecologically. They are the primary decomposers of dead and decaying organic matter, and help to cycle nutrients through the environment.

Fungi are widely used as a food source. Some are eaten directly. Some are used to help leaven bread, and others are used to ferment food products, such as wine and beer. Some fungi-derived enzymes are used in detergents, and some fungi are used as pesticides to control weeds, pests, and plant diseases.

Some fungi cause disease. For example, blight is a disease that affects potatoes. When the fungus infects the plants, the leaves turn brown. This makes it difficult for the plant to photosynthesize and grow. Athlete's foot is a fungal disease that affects people. It causes a rash between the toes and flaky, cracked skin. It is transmitted by touching infected surfaces of skin. It can be treated with antifungal medication.

Fungi and their underground world

Fungi aren't mobile, so they spread by making and releasing spores. The spores can germinate to form new fungi. Some fungi can also reproduce sexually.

The fruiting body of the mushroom contains tiny filaments called **hyphae**.

The hyphae extend into the ground and create a mass of tangled, interwoven threads. This underground network is called the **mycelium**.

As the underground hyphae grow, they absorb nutrients.

The filaments of mycorrhizae fungi form intimate associations with the roots of trees. The fungi supply the trees with water and nutrients, while the trees provide the fungi with the sugars they need for growth. This is an example of a symbiotic relationship.

Mycorrhizal networks often connect many different trees together. This is sometimes called the "wood wide web." Scientists think some plants use the wood wide web to communicate with one another. For example, if insects are attacking one tree, it may send a signal via the wood wide web that tells its neighbors to increase levels of insect-repelling chemicals. Trees also use the wood wide web to move sugars around. So, nutrients from one tree can nourish the cells of another tree.

Plants

The plant kingdom contains an estimated 320,000 species of plants. These include conifers, ferns, hornworts, liverworts, mosses, and flowering plants. Most plants are multicellular, but some are single-celled. Plants can reproduce sexually and asexually. Together, plants make up more than three-quarters of the world's biomass. They are the basis of most of the Earth's ecosystems.

Green plants use the sun's rays to help make food via photosynthesis. They lock up carbon and release oxygen as a by-product. The vast majority of the oxygen that we breathe comes from plants. Plants are widely used as a source of food. Wheat, barley, and lentils were some of the first plants to be domesticated many thousands of years ago.

Many animals have co-evolved with plants. Numerous insects, for example, pollinate flowers in exchange for pollen or nectar. Many animals disperse seeds when they eat them and then expel them in their feces. Some plants are carnivorous.

SPOTLIGHT ON PLANTS:
THE VENUS FLYTRAP
The Venus flytrap makes energy via photosynthesis, but it also catches organic food using modified leaves.

The trap only shuts if the hairs are touched twice within 10 seconds. After a further three touches, while the prey is struggling, the plant starts to produce digestive enzymes. This means that Venus flytrap plants can count. They estimate time and the number of touches.

Specialized trigger hairs detect the movement of insects and arachnids.

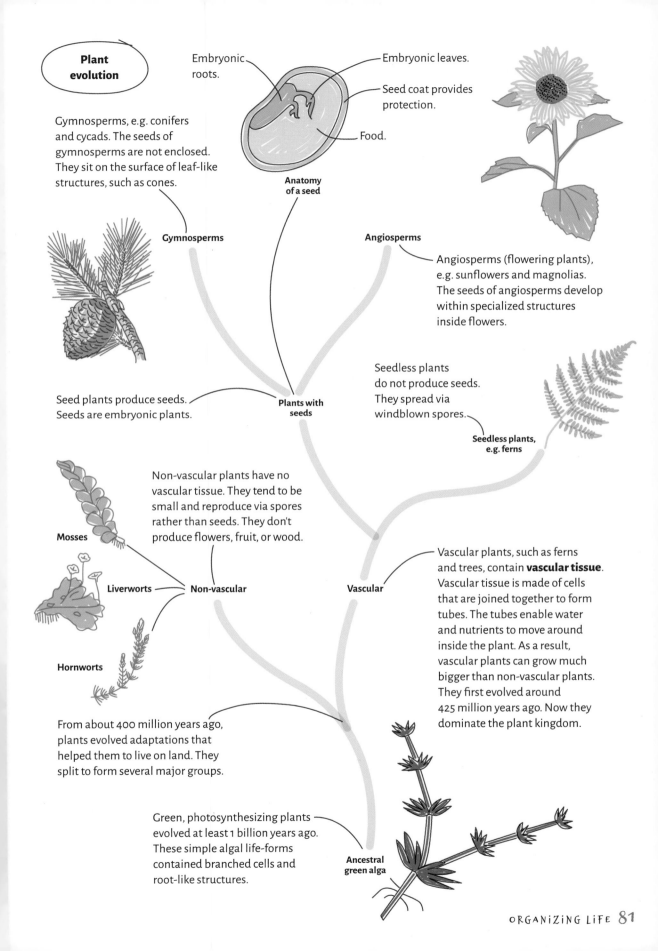

Plant evolution

Embryonic roots.

Embryonic leaves.

Seed coat provides protection.

Food.

Anatomy of a seed

Gymnosperms, e.g. conifers and cycads. The seeds of gymnosperms are not enclosed. They sit on the surface of leaf-like structures, such as cones.

Gymnosperms

Angiosperms

Angiosperms (flowering plants), e.g. sunflowers and magnolias. The seeds of angiosperms develop within specialized structures inside flowers.

Seed plants produce seeds. Seeds are embryonic plants.

Plants with seeds

Seedless plants do not produce seeds. They spread via windblown spores.

Seedless plants, e.g. ferns

Non-vascular plants have no vascular tissue. They tend to be small and reproduce via spores rather than seeds. They don't produce flowers, fruit, or wood.

Mosses

Liverworts

Non-vascular

Vascular

Vascular plants, such as ferns and trees, contain **vascular tissue**. Vascular tissue is made of cells that are joined together to form tubes. The tubes enable water and nutrients to move around inside the plant. As a result, vascular plants can grow much bigger than non-vascular plants. They first evolved around 425 million years ago. Now they dominate the plant kingdom.

Hornworts

From about 400 million years ago, plants evolved adaptations that helped them to live on land. They split to form several major groups.

Green, photosynthesizing plants evolved at least 1 billion years ago. These simple algal life-forms contained branched cells and root-like structures.

Ancestral green alga

Animals

Scientists have formally described around 1.5 million different animal species, of which 1 million are insects. Animals range from the microscopic to the massive. They form complex relationships with one another and with the environment.

Animals are multicellular eukaryotic organisms. They contain an enormous variety of specialized cells, such as nerve cells and muscle cells, which they use to help carry out complex functions, such as transmitting electrical information and movement.

Most animals consume organic material, breathe oxygen, move around, reproduce sexually, and develop from a tiny hollow ball of cells called a **blastula**.

Like fungi, animals are heterotrophs. They cannot make their own food, so they have to acquire nutrients externally. Unlike fungi, which absorb food, animals acquire food by eating and digesting other organisms.

Although the kingdom of animals is very diverse, its members share a relatively small number of body layouts.

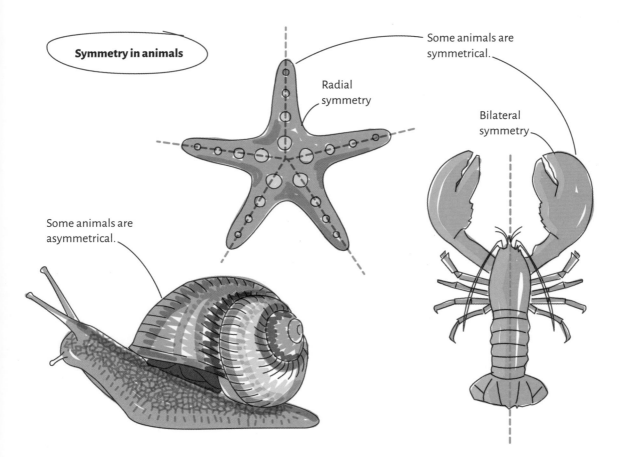

Symmetry in animals

Radial symmetry

Some animals are symmetrical.

Bilateral symmetry

Some animals are asymmetrical.

All snails are **asymmetrical**, e.g. garden snails have helical shells. Immobile animals, such as sea sponges, are also asymmetrical.

Some ocean-dwellers, such as starfish and sea urchins, are **radially symmetrical**. They show symmetry around a central axis.

Most species are **bilaterally symmetrical**. This means they have two almost identical halves, e.g. lobsters.

CLASSIFYING ANIMALS

Although animals are a very diverse group of organisms, they fall neatly into two categories. Animals can be divided into **vertebrates**, which have a backbone, and **invertebrates**, which don't have a backbone.

Vertebrates

Many of the animals we are familiar with, such as dogs, frogs, and fish, are vertebrates. There are around 70,000 different vertebrate species, but they make up less than 5% of all described animal species.

All vertebrates have a stiff spinal cord that runs the length of the body. The mouth is found at one end, and the anus is found at the other.

All vertebrates belong to the same phylum—Chordata. This can be subdivided into five different classes: amphibians, birds, fish, mammals, and reptiles.

Amphibians
* ★ Moist, permeable skin for breathing
* ★ Lay jelly eggs that hatch to produce fish-like larvae, e.g. frog eggs hatch into tadpoles
* ★ May produce poisonous chemicals for protection
* ★ Live in water and on land
* ★ **Cold-blooded**: body temperature depends on the surroundings

Birds
* ★ Feathers for warmth, waterproofing, and flight
* ★ Wings for flight
* ★ Beaks for eating, preening, and finding food
* ★ Lay hard-shelled eggs that are nearly always incubated by adult birds
* ★ **Warm-blooded**: keep a constant internal temperature

Fish
* ★ Gills for breathing
* ★ Scales for protection and streamlined shape
* ★ Air-filled swim bladder for buoyancy
* ★ Multiple fins for movement
* ★ Live in water
* ★ Cold-blooded

Mammals
* ★ Hair for warmth and camouflage
* ★ Specialized teeth, e.g. pointy front teeth for biting, and flatter back teeth for chewing
* ★ Babies usually develop inside their mother
* ★ Most mammals give birth to live young
* ★ Babies feed on milk from their mothers
* ★ Warm-blooded

Reptiles
* ★ Lungs for breathing
* ★ Waterproof skin that is shed during growth
* ★ Lay rubbery-shelled eggs that hatch to produce mini adults
* ★ Live in water or on land
* ★ Cold-blooded

Invertebrates

Most animals on Earth are invertebrates. Invertebrates are an incredibly successful and diverse group of animals. They have been around for more than 400 million years and have evolved to occupy just about every type of habitat, from oceans to land to sky.

Invertebrates have soft inner bodies that are held in shape by a flexible covering of outer cells or by a hard shell called an **exoskeleton**. There are more than 30 different phyla of invertebrates that we are more or less familiar with. They include: annelids, arthropods, cnidarians, echinoderms, and mollusks.

> **Some invertebrate phyla**

Annelids, e.g. earthworms and leeches
★ Long, segmented body
★ Tiny hairs stick out and help with movement
★ Live in water and on land

Arthropods, e.g. butterflies and spiders
★ Hard outer skeleton
★ Paired, jointed legs
★ Live in water and on land

Cnidarians, e.g. jellyfish and coral
★ Simple, sac-like bodies
★ Contain a single body cavity
★ Specialized stinging cells help to capture prey
★ Live in water

Echinoderms, e.g. starfish and brittle stars
★ Adult forms show radial symmetry
★ Hard, spiny covering or skin
★ Live in water

Mollusks, e.g. snails, slugs, and cephalopods, such as octopuses and squid
★ Unsegmented body
★ Muscular foot (and/or tentacles sometimes)
★ Toothed tongue called a radula
★ Live in water and on land

Invertebrate characteristics

Many invertebrates shed their outer skin layer as they grow. They are particularly vulnerable to predation during this time.

Many invertebrates, such as insects and crustaceans, hatch from eggs and turn into larvae that are different from the adult form. Sometimes, the larvae live in another environment. For example, hoverfly larvae live in the water, while hoverfly adults live on land. **Metamorphosis** occurs when the body form of the larva changes into its different adult form.

ARTHROPODS

The arthropods are a huge phylum of animals. They are divided into smaller groups called classes. These classes include insects, arachnids, crustaceans, and myriapods.

Most arthropods are relatively small, but the Japanese spider crab is enormous. It has the largest leg span of any arthropod. One specimen, caught 100 years ago, was heavier than a cocker spaniel dog and had an arm span longer than most cars.

(**Arthropod classes**)

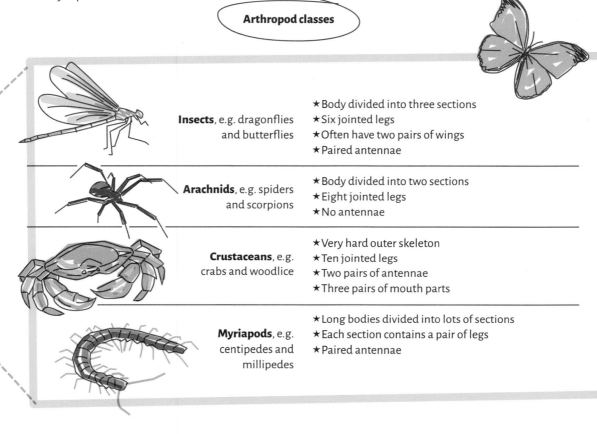

Insects, e.g. dragonflies and butterflies
★ Body divided into three sections
★ Six jointed legs
★ Often have two pairs of wings
★ Paired antennae

Arachnids, e.g. spiders and scorpions
★ Body divided into two sections
★ Eight jointed legs
★ No antennae

Crustaceans, e.g. crabs and woodlice
★ Very hard outer skeleton
★ Ten jointed legs
★ Two pairs of antennae
★ Three pairs of mouth parts

Myriapods, e.g. centipedes and millipedes
★ Long bodies divided into lots of sections
★ Each section contains a pair of legs
★ Paired antennae

Invertebrates eat a lot of different things, such as plants, insects, and crustaceans, so they have a variety of specialist mouthparts. For example, octopuses have bills, spiders have fangs, and beetles have slicing jaws.

Some invertebrates, such as aphids, can reproduce sexually and asexually. Queen honey bees, for example, produce an entire colony. Fertilized eggs hatch into female worker bees, while unfertilized eggs become male honey bee drones.

ViRUSES

Tiny infectious particles that can cause diseases are called **viruses**. They don't fit into the classification of life, because living things are made of cells, and viruses are not made from cells. Viruses are small packets of genetic material wrapped in a protein coat. They're important because they can infect all classes of living things, including animals, plants, and bacteria.

Living things reproduce, but viruses cannot reproduce on their own. Instead they must infect a cell and then hijack its internal machinery. The cell makes fresh copies of the virus, which are then released. Viruses are microscopic organisms that can be found just about everywhere on Earth.

Viruses tend to have regular shapes. They are very small. Most viruses are about one-hundredth the size of a bacterium.

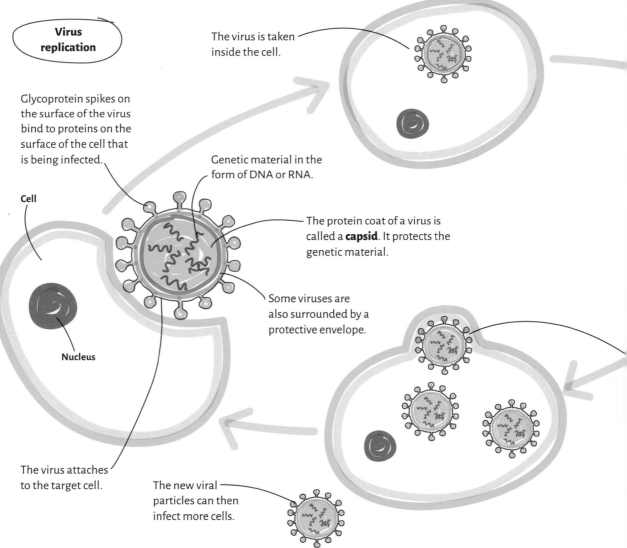

Virus replication

The virus is taken inside the cell.

Glycoprotein spikes on the surface of the virus bind to proteins on the surface of the cell that is being infected.

Genetic material in the form of DNA or RNA.

Cell

The protein coat of a virus is called a **capsid**. It protects the genetic material.

Nucleus

Some viruses are also surrounded by a protective envelope.

The virus attaches to the target cell.

The new viral particles can then infect more cells.

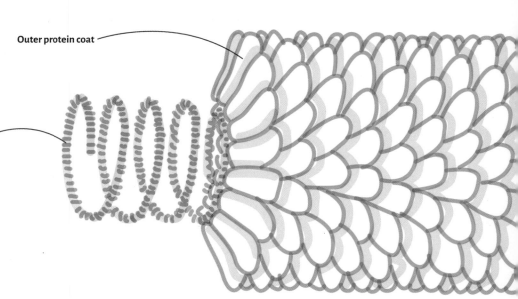

Outer protein coat

Coiled RNA core

The virus breaks apart. Genetic material from the virus travels to the nucleus of the cell where it is copied.

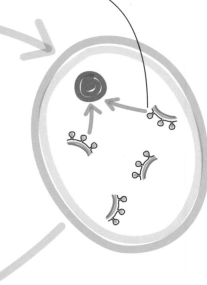

The cell makes new viral particles. The cell bursts and the new viral particles are released. The cell damage is what makes you feel ill.

Features of viruses

Most viruses are **pathogenic**. This means they cause disease. Viral diseases range in severity. Coronaviruses, for example, are a large family of viruses. They cause illness ranging from the common cold to more serious diseases such as Severe Acute Respiratory Syndrome (SARS) and COVID-19.

Some viruses are **zoonotic**. This means that sometimes, they can be transmitted from animals to people. It's thought that the global COVID-19 pandemic of 2020 began after a coronavirus found in bats "jumped" to a second species—possibly a pangolin— and then to humans.

Viruses have **structural adaptations** that enable them to infect living things. For example, coronaviruses are covered in tiny spikes that interact with key proteins on human cells. This enables the virus to latch on to and then enter a cell.

Viruses also infect plants. The tobacco mosaic virus infects many species of plants, including tobacco plants and tomatoes. It discolors the leaves and causes a mottled "mosaic" pattern. This makes it difficult for the plant to photosynthesize, so the plant's growth is affected.

Just like living organisms, viruses evolve. Mistakes are made as host cells copy the virus's genetic information. This leads to mutations. Sometimes the mutations have no effect. Sometimes they make the virus less infectious, but sometimes they make it more dangerous. For example, they might make the virus more infectious or give it the ability to jump between species.

Proposed by Carl Woese. Separates life into bacteria, archaea, and eukaryotes.

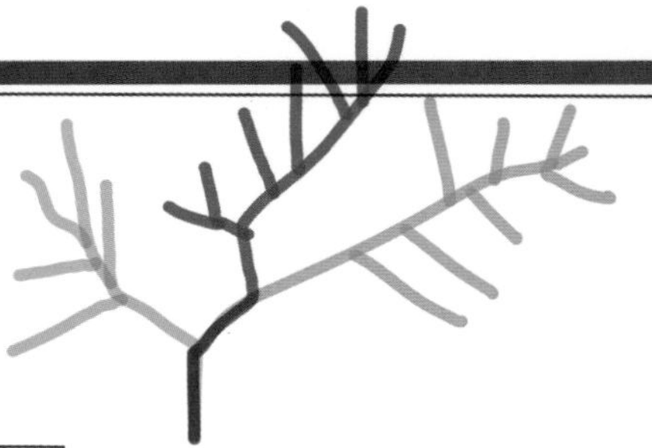

The grouping of life into organized categories. Helps to make sense of life.

THREE-DOMAIN SYSTEM

TAXONOMY

LINNAEAN SYSTEM

Proposed by Carl Linnaeus. Divides life into kingdoms, phyla, classes, orders, families, genera, and species.

CLASSIFICATION

ORGANIZING LIFE

Cause disease in animals, plants, and other organisms.

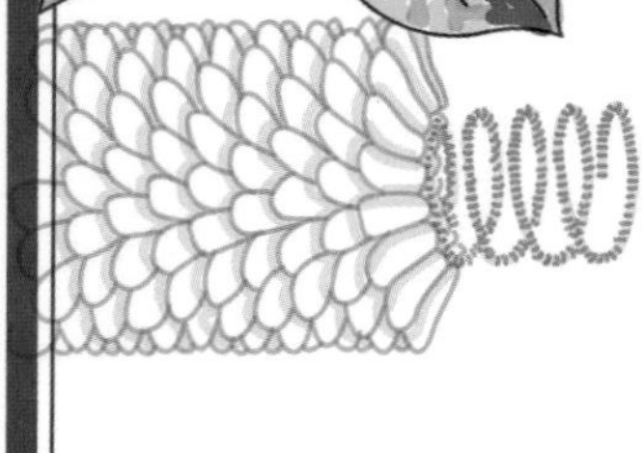

EVOLVE

Viruses change over time. Mutations occur as the virus is copied by living cells.

PATHOGENIC

VIRUSES

ZOONOTIC

Some viruses can jump from animals to humans, e.g. some coronaviruses.

NOT ALIVE OR DEAD

Defy classification. Can only reproduce inside the cells of living organisms.

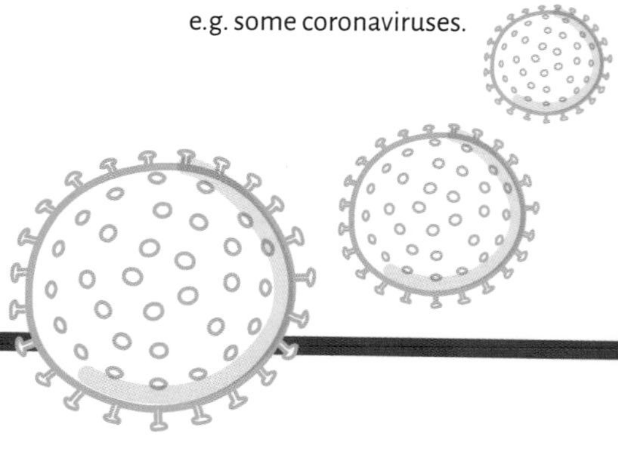

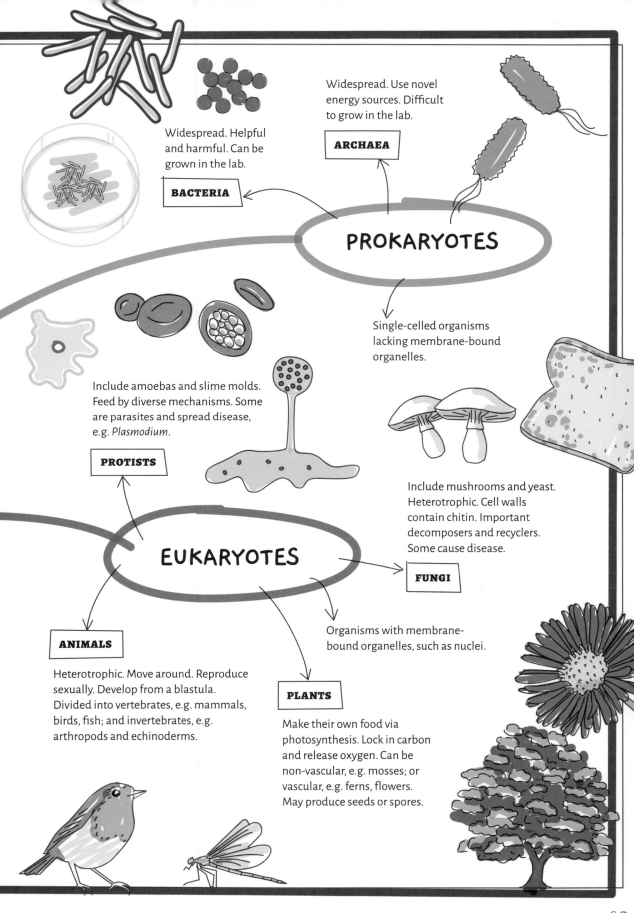

Widespread. Use novel
energy sources. Difficult
to grow in the lab.

ARCHAEA

Widespread. Helpful
and harmful. Can be
grown in the lab.

BACTERIA

PROKARYOTES

Single-celled organisms
lacking membrane-bound
organelles.

Include amoebas and slime molds.
Feed by diverse mechanisms. Some
are parasites and spread disease,
e.g. *Plasmodium*.

PROTISTS

Include mushrooms and yeast.
Heterotrophic. Cell walls
contain chitin. Important
decomposers and recyclers.
Some cause disease.

EUKARYOTES

FUNGI

Organisms with membrane-
bound organelles, such as nuclei.

ANIMALS

Heterotrophic. Move around. Reproduce
sexually. Develop from a blastula.
Divided into vertebrates, e.g. mammals,
birds, fish; and invertebrates, e.g.
arthropods and echinoderms.

PLANTS

Make their own food via
photosynthesis. Lock in carbon
and release oxygen. Can be
non-vascular, e.g. mosses; or
vascular, e.g. ferns, flowers.
May produce seeds or spores.

CHAPTER 6

METABOLISM

At any point in time, the cells that make up living things are busy carrying out thousands of vitally important chemical reactions. Metabolism is the term used to describe all of the chemical reactions that occur inside an organism. The reactions are controlled by proteins called enzymes. Many metabolic reactions, like the ones underpinning energy production, are shared between different groups of organisms, while others are more restricted. Photosynthesis, for example, is found in green plants, algae, and certain bacteria. In this chapter, you'll learn more about the metabolic reactions that keep life going.

CHEMICAL REACTIONS AND PATHWAYS

Cells are busy places. In order for them to keep functioning, there are many different chemical reactions happening all the time. Often, individual reactions are linked together to form bigger reactions.

When chemical reactions are joined together, they form more complex chemical pathways. Proteins called **enzymes** control the various steps along these pathways.

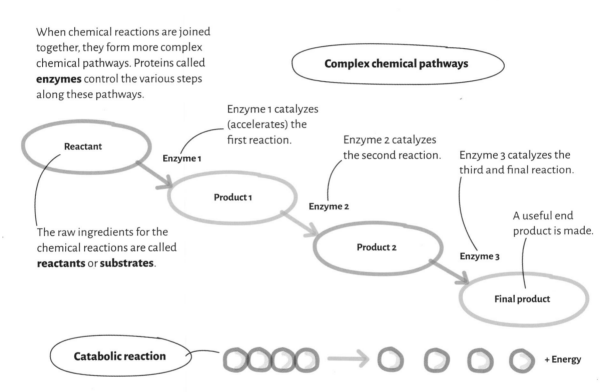

Complex chemical pathways

Reactant

Enzyme 1

Enzyme 1 catalyzes (accelerates) the first reaction.

The raw ingredients for the chemical reactions are called **reactants** or **substrates**.

Product 1

Enzyme 2

Enzyme 2 catalyzes the second reaction.

Product 2

Enzyme 3

Enzyme 3 catalyzes the third and final reaction.

A useful end product is made.

Final product

Catabolic reaction — + Energy

Metabolic reactions can be either catabolic or anabolic. **Catabolic** reactions break down larger molecules into smaller ones. They release energy. Respiration is a catabolic process that breaks down glucose. Excess proteins are also broken down catabolically. The product, urea, is then excreted in urine.

Anabolic reactions build bigger molecules from smaller ones. They require energy. Photosynthesis is an anabolic pathway. When glucose molecules are joined together to form carbohydrates, such as starch and cellulose, this is another anabolic reaction.

Anabolic reactions are also called **biosynthetic reactions** because they help to build biologically useful molecules. For example, cells combine glucose with nitrate ions to make amino acids, which are then used to build proteins. Proteins are biologically useful because they play important roles, such as helping cells to communicate and controlling cellular reactions.

Anabolic reaction — + Energy —

ENZYMES

Enzymes are biological **catalysts**. This means that they speed up chemical reactions, but they are not changed by the reaction themselves. There are thousands of different enzymes. Each one interacts with a different substrate. Enzymes are protein molecules that are intricately folded into precise, 3D shapes. This allows smaller molecules, called the substrates, to bind to them.

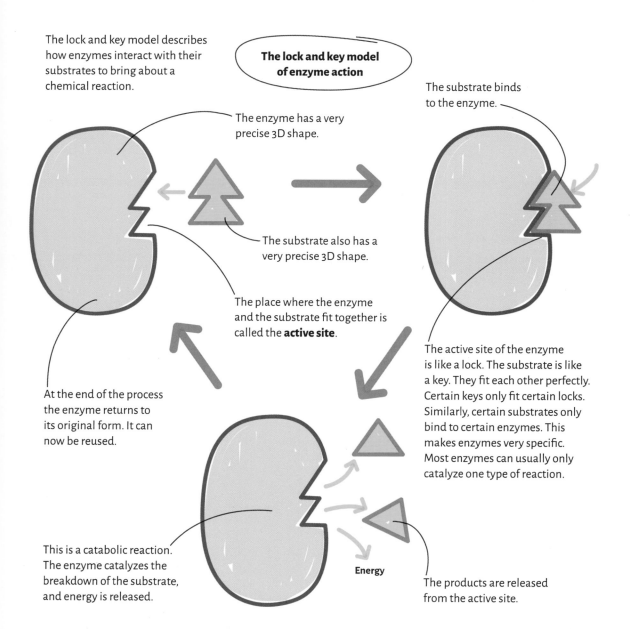

The lock and key model describes how enzymes interact with their substrates to bring about a chemical reaction.

The lock and key model of enzyme action

The enzyme has a very precise 3D shape.

The substrate also has a very precise 3D shape.

The place where the enzyme and the substrate fit together is called the **active site**.

The substrate binds to the enzyme.

The active site of the enzyme is like a lock. The substrate is like a key. They fit each other perfectly. Certain keys only fit certain locks. Similarly, certain substrates only bind to certain enzymes. This makes enzymes very specific. Most enzymes can usually only catalyze one type of reaction.

At the end of the process the enzyme returns to its original form. It can now be reused.

This is a catabolic reaction. The enzyme catalyzes the breakdown of the substrate, and energy is released.

Energy

The products are released from the active site.

Factors affecting metabolic speed

Metabolic reactions don't always proceed at the same rate. Their speed can be influenced by many different factors. Substrate concentration is one of them.

As the number of substrate molecules increases, there are more keys available to fit the locks. This speeds up the reaction, but only to a point.

When all of the active sites are occupied, there are no more locks for the keys to fit. The enzyme is said to be **saturated**. At this point, adding more substrate to the enzyme has no effect on the speed of the reaction.

Temperature also has an effect on the speed of metabolic reactions.

As temperature increases, the rate of the reaction increases.

The reaction has an **optimum temperature**. This is when the rate is at its maximum.

Above the optimum temperature, the rate of the reaction declines. This is because extreme heat can cause the shape of the active site to change. If this happens, the substrate can no longer bind to the active site. The reaction slows down or stops and the enzyme is said to be **denatured**.

Enzymes are also affected by the **pH** in the local environment. The pH is a measure of how acidic or alkaline something is. Each enzyme works best at a specific or optimum pH. This is influenced by the place in the body where the enzyme works.

In the stomach, enzymes work best at acidic pH values.

In most human cells, enzymes work best at neutral pH of around 7.

If the pH strays too far from the optimum value, the rate of the reaction decreases. This is because the shape of the active site changes and the substrate can no longer bind. Once again, the enzyme is denatured.

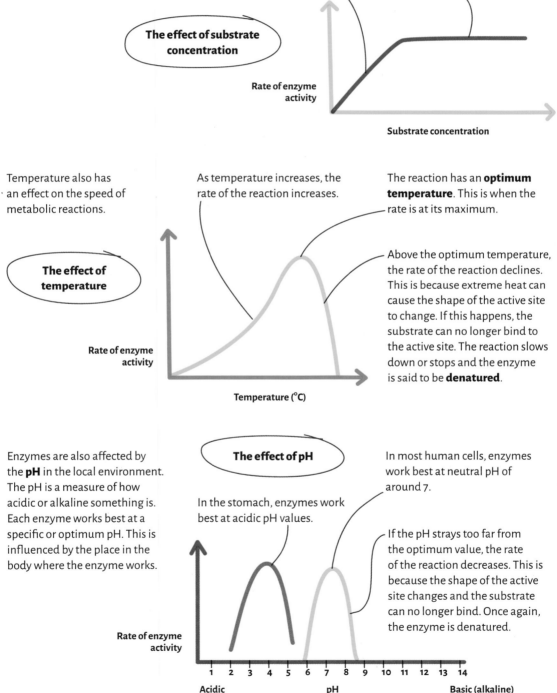

The effect of substrate concentration

Rate of enzyme activity

Substrate concentration

The effect of temperature

Rate of enzyme activity

Temperature (°C)

The effect of pH

Rate of enzyme activity

1 2 3 4 5 6 7 8 9 10 11 12 13 14

Acidic   pH   Basic (alkaline)

METABOLIC RATE

An organism's metabolic rate is the amount of energy that it uses in a given amount of time—usually 24 hours. It is measured in joules (J), calories (cal), or kilocalories (kcal). One kilocalorie is the same as 1,000 calories or approximately 4,200 joules.

Metabolic rate depends on activity. When an organism is resting or asleep, metabolic rate is low. This is called the **basal metabolic rate**. During this time, the body needs comparatively little energy; just enough to keep vital organs like the heart, lungs, and brain functioning properly.

Metabolic rate can be measured in different ways, including oxygen consumption, carbon dioxide production, and heat production.

Calorimeters are instruments that are used to measure metabolic rate. Sophisticated laboratory calorimeters can be used to calculate the amount of energy in food. Simple calorimeters like the one shown here are relatively easy to make.

Energy in food (joules per gram) = Mass of water (grams) x Temperature rise (°C) x 4.2 / Mass of food (grams)

If 1 g of food is burned, and 10 cm³ of water warms up by 15°C:

10 cm³ of water has a mass of 10 g

Energy content of food = 10 x 15 x 4.2 / 1 = 630 J/g

Measuring metabolic rate

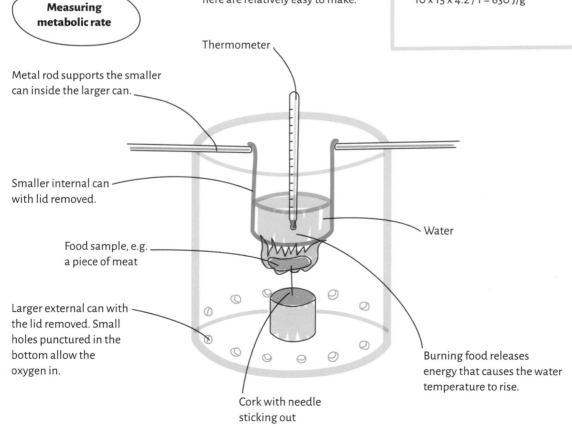

Thermometer

Metal rod supports the smaller can inside the larger can.

Smaller internal can with lid removed.

Food sample, e.g. a piece of meat

Larger external can with the lid removed. Small holes punctured in the bottom allow the oxygen in.

Water

Burning food releases energy that causes the water temperature to rise.

Cork with needle sticking out

Energy requirements

Different organisms and individuals have different energy requirements and different metabolic rates.

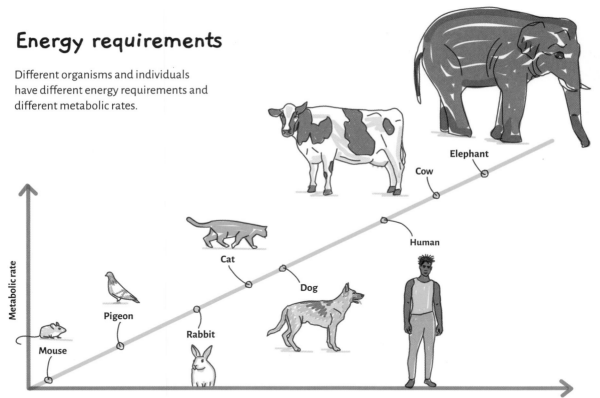

Generally speaking, organisms with more mass have higher metabolic rates than smaller organisms with less mass, e.g. elephants have higher metabolic rates than mice. This makes sense because bigger animals have more cells and so need more energy to power those cells. We say that body size is directly proportional to metabolic rate.

Animals with higher metabolic rates need to be able to deliver oxygen efficiently to their cells. Birds and mammals have higher metabolic rates than reptiles and amphibians, which in turn have higher rates than fish. These requirements are reflected in their anatomy.

Anatomy reflects metabolic requirements

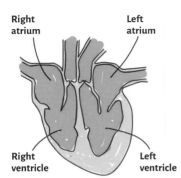

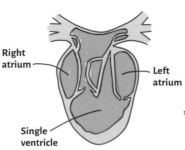

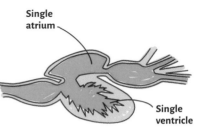

Mammals and birds have a four-chambered heart and a complex circulatory system that separates oxygenated and deoxygenated blood.

Amphibians and most reptiles have a three-chambered heart, which limits the mixing of oxygenated and deoxygenated blood.

Fish have a two-chambered heart. It pumps deoxygenated blood to the gills where it is oxygenated and then delivered to the rest of the body.

CONTROL OF METABOLISM

Molecules called **inhibitors** can block the action of enzymes and so influence metabolism. Pharmaceutical manufacturers often utilize these molecules to make new drugs.

Living things possess many naturally occurring enzyme inhibitors, but many human-made drugs also work in this way.

Antibiotics, for example, work by inhibiting bacterial enzymes. Penicillin kills bacteria by blocking the active site of the enzyme that is used by bacteria to build cell walls.

Toxins, such as sarin, mercury, and cyanide, are also enzyme inhibitors. Sarin binds to and blocks an active site on an enzyme called acetylcholinesterase, which helps to control how nerve cells work.

Most enzyme inhibitors are not harmful. They are a normal part of metabolism and help cellular processes in the body to run smoothly. The body's many metabolic pathways are influenced by genetic mechanisms. Key genes determine when and where certain enzymes are active.

Some metabolic pathways are unique to certain groups of organisms. For example, Antarctic toothfish contain metabolic pathways that lead to the production of anti-freeze proteins. These proteins help the fish survive at subzero temperatures.

Other metabolic pathways are more widespread, for example, **glycolysis** is part of respiration. It is the first step in a complex metabolic pathway that converts sugar from food into energy. It involves 10 enzyme-catalyzed reactions that are the same in all animals, plants, and bacteria. This is because glycolysis evolved a long time ago in the common ancestor of these organisms.

Antarctic toothfish

Enzyme inhibitors

There are different types of inhibitors that work in different ways. **Competitive inhibition** occurs when an inhibitor binds to an enzyme's active site and prevents the usual substrate from binding.

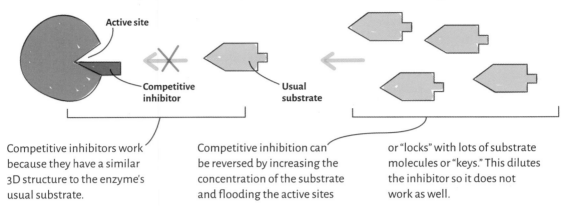

Active site
Competitive inhibitor
Usual substrate

Competitive inhibitors work because they have a similar 3D structure to the enzyme's usual substrate.

Competitive inhibition can be reversed by increasing the concentration of the substrate and flooding the active sites

or "locks" with lots of substrate molecules or "keys." This dilutes the inhibitor so it does not work as well.

Noncompetitive inhibition occurs when an inhibitor binds to part of the enzyme that is not the active site.

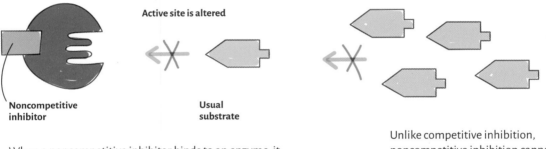

Active site is altered

Noncompetitive inhibitor

Usual substrate

When a noncompetitive inhibitor binds to an enzyme, it changes the shape of the active site so the usual substrate can no longer bind. As a result, the reaction slows down.

Unlike competitive inhibition, noncompetitive inhibition cannot be reversed by increasing the concentration of the substrate.

Feedback inhibition occurs when the final product in a metabolic pathway "feeds back" and binds to an enzyme at the start of the same pathway. This closes down the pathway.

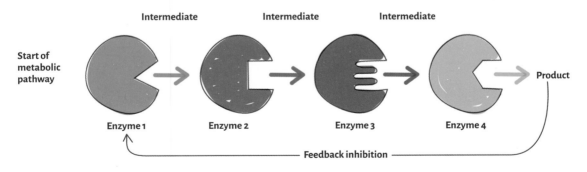

Intermediate   Intermediate   Intermediate

Start of metabolic pathway

Enzyme 1   Enzyme 2   Enzyme 3   Enzyme 4   Product

Feedback inhibition

Feedback inhibition is a form of **negative feedback**, whereby a single event leads to a decrease in overall function. This is an example of **homeostasis**, where biological systems self-regulate to maintain the conditions that are optimal for survival.

Feedback inhibition is reversible. When the concentration of the inhibiting molecule falls, the enzyme is able to become active again. The reaction is catalyzed and starts over again.

RESPIRATION

espiration is a major metabolic pathway. It involves a variety of steps and many different enzymes, and it takes place inside the cells of all living things. Respiration is important because it liberates energy from food, providing organisms with the fuel needed for important processes, such as growth, repair, and movement.

Aerobic respiration

Aerobic respiration involves oxygen. During aerobic respiration, glucose reacts with oxygen to produce carbon dioxide, water, and energy.

The raw materials for aerobic respiration come from the environment. Oxygen is in the air we breathe. Glucose is in the food we eat.

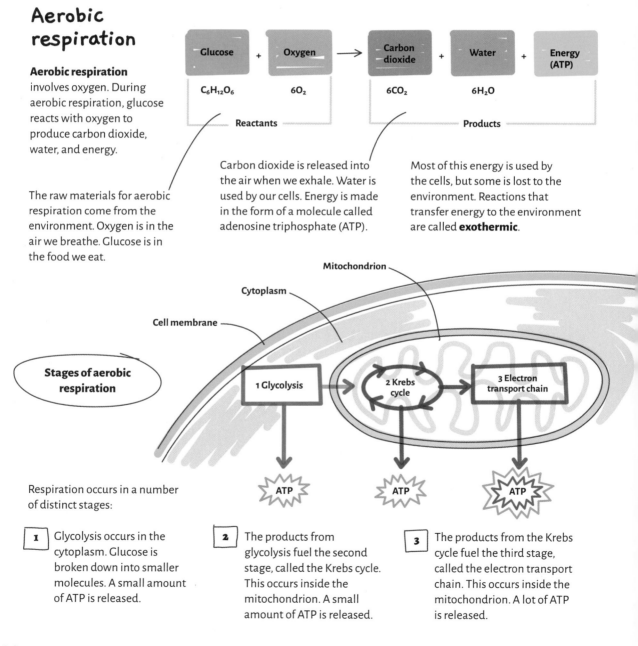

Glucose	+	Oxygen	→	Carbon dioxide	+	Water	+	Energy (ATP)
$C_6H_{12}O_6$		$6O_2$		$6CO_2$		$6H_2O$		

Reactants — Products

Carbon dioxide is released into the air when we exhale. Water is used by our cells. Energy is made in the form of a molecule called adenosine triphosphate (ATP).

Most of this energy is used by the cells, but some is lost to the environment. Reactions that transfer energy to the environment are called **exothermic**.

Mitochondrion

Cytoplasm

Cell membrane

Stages of aerobic respiration

1 Glycolysis → 2 Krebs cycle → 3 Electron transport chain

ATP   ATP   ATP

Respiration occurs in a number of distinct stages:

1 | Glycolysis occurs in the cytoplasm. Glucose is broken down into smaller molecules. A small amount of ATP is released.

2 | The products from glycolysis fuel the second stage, called the Krebs cycle. This occurs inside the mitochondrion. A small amount of ATP is released.

3 | The products from the Krebs cycle fuel the third stage, called the electron transport chain. This occurs inside the mitochondrion. A lot of ATP is released.

Anaerobic respiration

Anaerobic respiration does not involve oxygen. During anaerobic respiration, glucose is broken down to produce various products including energy.

Most living things respire aerobically, but when oxygen is scarce, some organisms start to respire anaerobically, e.g. our muscles respire anaerobically if they lack oxygen.

Anaerobic respiration has two big drawbacks:

1. Lactic acid builds up in the muscles, causing pain and fatigue. Excess is broken down in the liver. This requires oxygen. The amount of oxygen needed to break lactic acid down is known as the **oxygen debt**. This explains why people puff and pant after stopping exercise. It is the body's way of repaying the oxygen debt.

2. Glucose is not fully broken down. This makes anaerobic respiration less efficient than aerobic respiration. Less energy is produced per molecule of glucose used.

Anaerobic respiration in animals

Glucose → Lactic acid + Energy

In animals, glucose is broken down to produce lactic acid and energy.

Anaerobic respiration in plants and microbes

Glucose → Ethanol + Carbon dioxide + Energy

Anaerobic respiration also occurs in plants. Here, glucose is broken down to produce ethanol, carbon dioxide, and energy.

Microbes can respire anaerobically too. Some produce lactic acid. Others produce ethanol and carbon dioxide. When microbes respire anaerobically, it is called **fermentation**.

Fermentation is economically important. Yeast, for example, is used to make beer. Anaerobically respiring yeast cells produce ethanol, which makes the beer alcoholic, and carbon dioxide, which makes it fizzy.

Yeast is also used to make bread. Anaerobic respiration is fueled by sugar in the dough, and the carbon dioxide that is produced makes the bread rise. Alcohol is also produced, but it evaporates as the bread bakes.

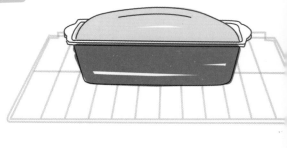

PHOTOSYNTHESIS

Another major metabolic pathway is called **photosynthesis.**
It is the process that plants use to make food. Oxygen is also
produced as a by-product. Plants have a variety of special
adaptations that help them to carry out photosynthesis.

How photosynthesis works

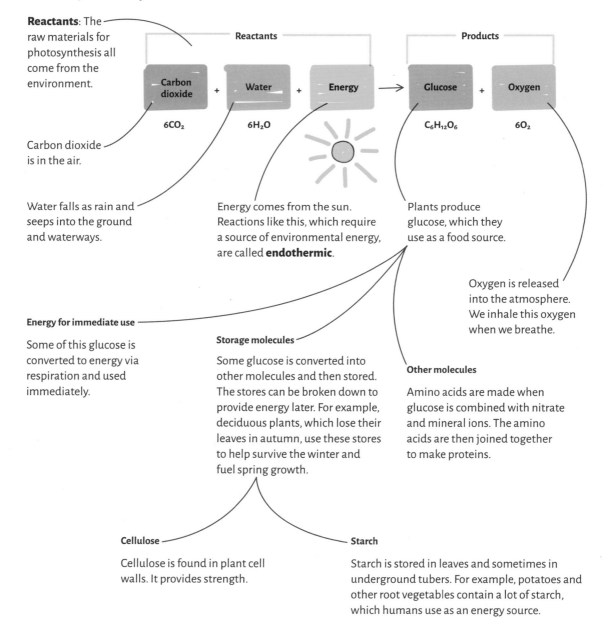

Reactants: The raw materials for photosynthesis all come from the environment.

Carbon dioxide is in the air.

Water falls as rain and seeps into the ground and waterways.

Reactants

$6CO_2$ + $6H_2O$ + Energy → Glucose + Oxygen

Carbon dioxide

Water

Energy

Products

Glucose

$C_6H_{12}O_6$

Oxygen

$6O_2$

Energy comes from the sun. Reactions like this, which require a source of environmental energy, are called **endothermic**.

Plants produce glucose, which they use as a food source.

Oxygen is released into the atmosphere. We inhale this oxygen when we breathe.

Energy for immediate use

Some of this glucose is converted to energy via respiration and used immediately.

Storage molecules

Some glucose is converted into other molecules and then stored. The stores can be broken down to provide energy later. For example, deciduous plants, which lose their leaves in autumn, use these stores to help survive the winter and fuel spring growth.

Other molecules

Amino acids are made when glucose is combined with nitrate and mineral ions. The amino acids are then joined together to make proteins.

Cellulose

Cellulose is found in plant cell walls. It provides strength.

Starch

Starch is stored in leaves and sometimes in underground tubers. For example, potatoes and other root vegetables contain a lot of starch, which humans use as an energy source.

The relationship between photosynthesis and respiration

Photosynthesis and respiration are intimately connected. This relationship enables life on Earth to survive. The products of photosynthesis are the reactants for respiration, and the products of respiration are the reactants for photosynthesis. The equation for respiration is the direct opposite of the equation for photosynthesis.

Chloroplasts are specialized organelles that are found inside the cells of green plants. This is where photosynthesis occurs.

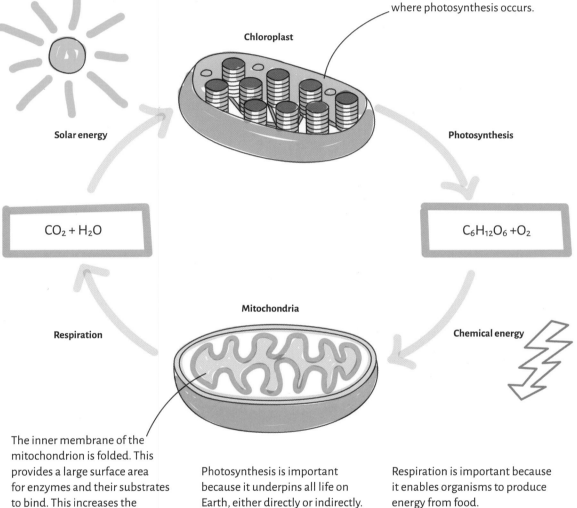

Chloroplast

Solar energy

Photosynthesis

$$CO_2 + H_2O$$

$$C_6H_{12}O_6 + O_2$$

Mitochondria

Respiration

Chemical energy

The inner membrane of the mitochondrion is folded. This provides a large surface area for enzymes and their substrates to bind. This increases the efficiency of respiration.

Photosynthesis is important because it underpins all life on Earth, either directly or indirectly. Plants make up most of the Earth's biomass. When they are consumed, the nutrients pass up the food chain. Photosynthesis produces an estimated 150 billion metric tonnes (165 billion US tons) of carbohydrates per year. We use this for food, and use fertilizers to boost plant growth.

Respiration is important because it enables organisms to produce energy from food.

Respiration and photosynthesis are also important because they form part of the **carbon cycle**. The carbon cycle is the set of pathways used to recycle carbon in the environment.

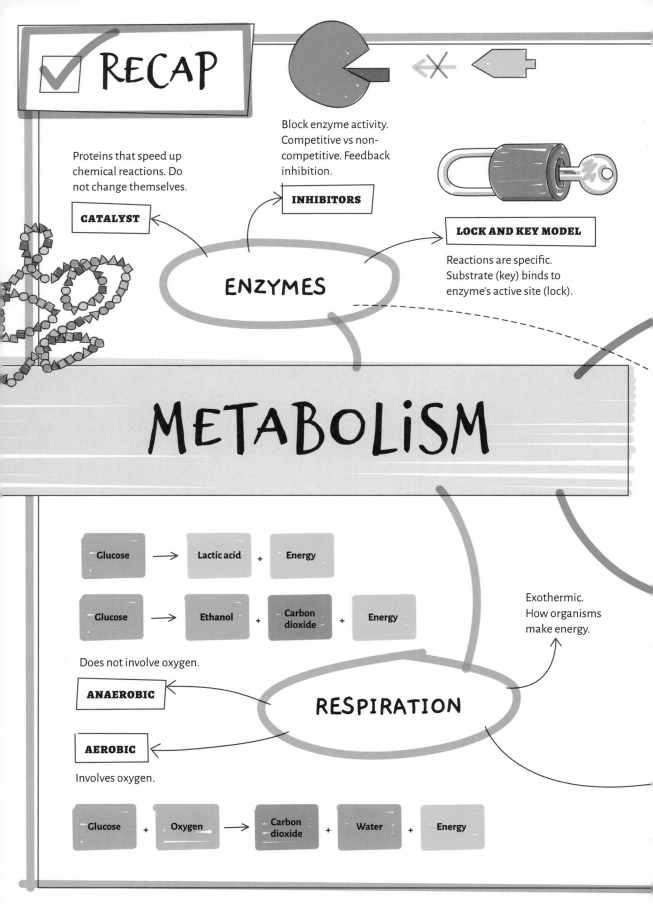

Proteins that speed up
chemical reactions. Do
not change themselves.

CATALYST

Block enzyme activity.
Competitive vs non-
competitive. Feedback
inhibition.

INHIBITORS

LOCK AND KEY MODEL

Reactions are specific.
Substrate (key) binds to
enzyme's active site (lock).

ENZYMES

METABOLISM

| Glucose | → | Lactic acid | + | Energy |

| Glucose | → | Ethanol | + | Carbon dioxide | + | Energy |

Does not involve oxygen.

ANAEROBIC

AEROBIC

Involves oxygen.

RESPIRATION

Exothermic.
How organisms
make energy.

| Glucose | + | Oxygen | → | Carbon dioxide | + | Water | + | Energy |

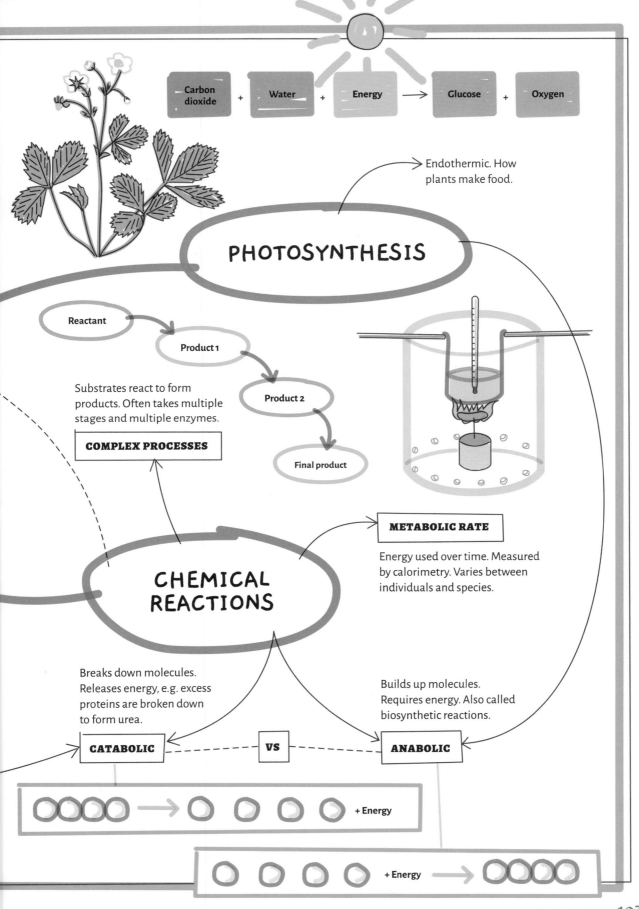

Carbon dioxide + Water + Energy → Glucose + Oxygen

Endothermic. How plants make food.

PHOTOSYNTHESIS

Reactant

Product 1

Product 2

Final product

Substrates react to form products. Often takes multiple stages and multiple enzymes.

COMPLEX PROCESSES

METABOLIC RATE

Energy used over time. Measured by calorimetry. Varies between individuals and species.

CHEMICAL REACTIONS

Breaks down molecules. Releases energy, e.g. excess proteins are broken down to form urea.

Builds up molecules. Requires energy. Also called biosynthetic reactions.

CATABOLIC — — — **VS** — — — **ANABOLIC**

+ Energy

+ Energy →

CHAPTER 7

PLANT STRUCTURE AND FUNCTION

Plants are responsive, organized, coordinated living things. Just like animals, they have cells arranged into tissues and organs that work together to form productive systems. Plants are highly specialized, so they can acquire water and minerals from the soil and produce food from sunlight via photosynthesis. Their leaves contain a complex arrangement of cells, and their vascular system is specialized to transport water and food. Plants can grow upward and outward, and much of their behavior is coordinated by hormones. They can suffer from diseases and deficiencies, but they have evolved a sophisticated suite of adaptations to maintain health. Let's find out more about plants.

TRANSPIRATION

Plants need a continuous supply of water to power photosynthesis and help them grow. This process is called **transpiration**. Water enters the plant via the roots. It is then drawn up into the stems and leaves and evaporates into the atmosphere.

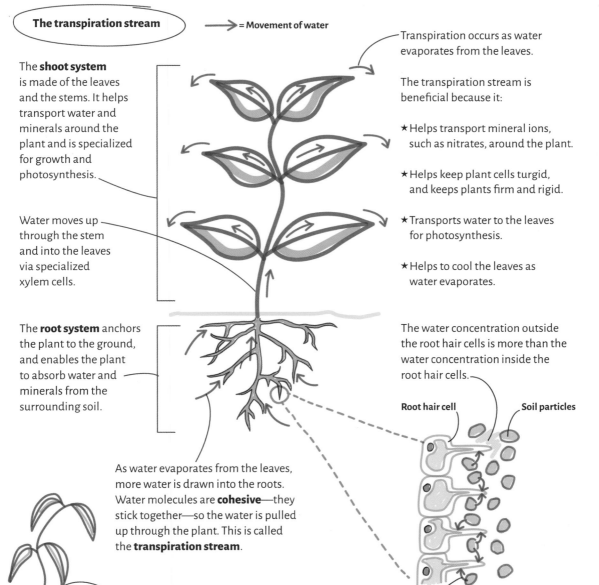

The transpiration stream

⟶ = Movement of water

The **shoot system** is made of the leaves and the stems. It helps transport water and minerals around the plant and is specialized for growth and photosynthesis.

Water moves up through the stem and into the leaves via specialized xylem cells.

The **root system** anchors the plant to the ground, and enables the plant to absorb water and minerals from the surrounding soil.

Transpiration occurs as water evaporates from the leaves.

The transpiration stream is beneficial because it:

★ Helps transport mineral ions, such as nitrates, around the plant.

★ Helps keep plant cells turgid, and keeps plants firm and rigid.

★ Transports water to the leaves for photosynthesis.

★ Helps to cool the leaves as water evaporates.

The water concentration outside the root hair cells is more than the water concentration inside the root hair cells.

Root hair cell   **Soil particles**

As water evaporates from the leaves, more water is drawn into the roots. Water molecules are **cohesive**—they stick together—so the water is pulled up through the plant. This is called the **transpiration stream**.

If the plant does not absorb enough water, the cell vacuoles will shrivel and the plant will wilt. Without water, photosynthesis stops and the plant may die.

Water moves into the root hairs via osmosis.

THE VASCULAR SYSTEM

Just like animals, plants have a vascular system that helps them to move key molecules around. Instead of arteries and vessels, plants have a series of hollow tubes made from specialized cells. There are two different types of vessels: xylem and phloem.

Xylem vessel

Xylem vessels transport water and mineral ions from the roots to the stems and the leaves.

When they are first formed, xylem vessels are made of living cells, but over time, a chemical called lignin forms hollow spirals inside these cells. The cells die and leave behind a hollow tube fortified with lignin scaffolding.

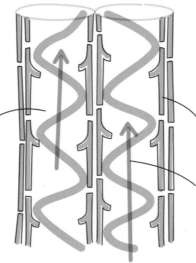

Lignified cells are also known as "wood." The lignin spirals are very strong. They provide structural support and help the plant to stand up.

Cell wall

Via transpiration, water and minerals flow in one direction only—upward.

Phloem vessel

Phloem vessels transport the sugars made by photosynthesis to other parts of the plant for immediate use (e.g. in the meristems) or for storage (e.g. bulbs and tubers).

Phloem cell—an elongated, living cell with pores in the end walls. The pores allow water and dissolved substances to pass along the vessel, from one cell to another. To make more room for the nutrients to flow, phloem cells lack many basic organelles, including nuclei and ribosomes.

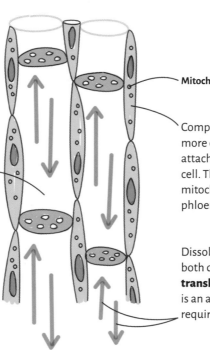

Mitochondria

Companion cell—one or more companion cells are attached to each phloem cell. They have lots of mitochondria and provide the phloem cells with energy.

Dissolved sugars can flow in both directions. This is called **translocation**. Translocation is an active process, so it requires energy.

PLANT GROWTH

Unlike most animals, plants continue to grow throughout their lives. They can do this because they have specialized growth zones called **meristems**. Meristems contain actively dividing stem cells that generate new plant tissue.

Primary growth

When roots and shoots grow longer, and when plants grow taller, it is called **primary growth**. Primary growth occurs because there are meristems at the tips of the roots and shoots. They are called **apical meristems**.

Some of the new cells are stem cells. They remain in the meristem and keep producing more new cells.

Root hair cells

Some of the new cells migrate away from the meristem and differentiate to form specialized cell types, such as root hair cells and epidermal cells

Zone of cell division

Apical meristem: Stem cells here divide by mitosis to produce new cells. This is primary growth.

End of root

Secondary growth

Some plants, such as trees, also grow in circumference. This is due to **secondary growth**. Secondary growth occurs because the stems of these plants contain additional meristems. These **lateral meristems** are arranged in circular rows of dividing cells.

As new cells are made, the stem or trunk increases in diameter. Secondary growth produces the tree rings that are sometimes used to determine their age.

The **lateral meristem** contains stem cells that divide by mitosis to produce new cells. This is secondary growth.

Newer growth

Old growth

Bark

LEAF STRUCTURE

Plant organs are also highly specialized. Leaves are specially adapted to maximize photosynthesis and the transport of substances into and out of the plant. They contain a variety of tissue types, and their cells contain specialized organelles called **chloroplasts**. This is where photosynthesis occurs. Chloroplasts contain the green pigment **chlorophyll**, which absorbs energy from sunlight.

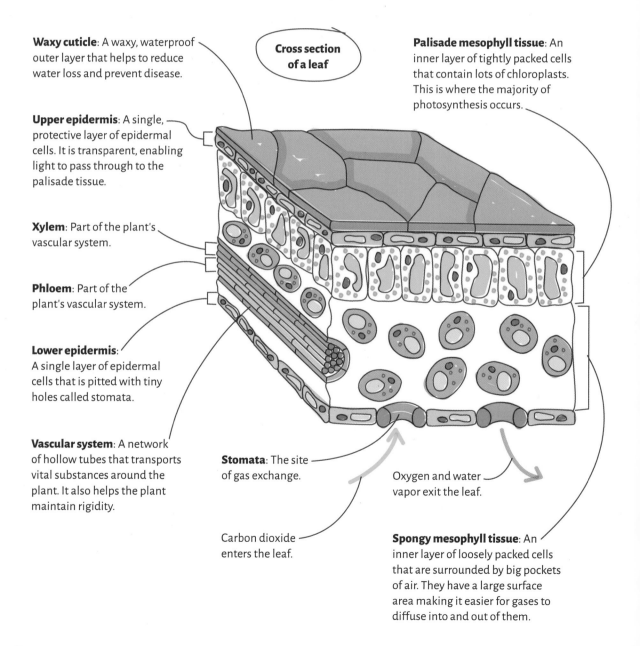

Cross section of a leaf

Waxy cuticle: A waxy, waterproof outer layer that helps to reduce water loss and prevent disease.

Upper epidermis: A single, protective layer of epidermal cells. It is transparent, enabling light to pass through to the palisade tissue.

Xylem: Part of the plant's vascular system.

Phloem: Part of the plant's vascular system.

Lower epidermis: A single layer of epidermal cells that is pitted with tiny holes called stomata.

Vascular system: A network of hollow tubes that transports vital substances around the plant. It also helps the plant maintain rigidity.

Palisade mesophyll tissue: An inner layer of tightly packed cells that contain lots of chloroplasts. This is where the majority of photosynthesis occurs.

Stomata: The site of gas exchange.

Oxygen and water vapor exit the leaf.

Carbon dioxide enters the leaf.

Spongy mesophyll tissue: An inner layer of loosely packed cells that are surrounded by big pockets of air. They have a large surface area making it easier for gases to diffuse into and out of them.

Gas exchange

Gases enter and exit the leaves via stomata. **Stomata** are tiny holes found in the epidermis of leaves, stems, and other plant organs. Plants need to obtain carbon dioxide for photosynthesis, and then to expel the waste oxygen that is produced. The stomata help with this process.

Transpiration is affected by environmental conditions, such as temperature, humidity, and wind. Plants react to these changes by opening and closing their stomata, so stomata also help to control transpiration.

Stomata

Most plants open their stomata during the day to help photosynthesis. If it is particularly dry or hot, a plant may close its stomata during the day to prevent excess water loss.

Daytime

Guard cells are specialized cells that surround the stomata. They control the size and opening of the pore.

When the guard cells are swollen, the stomata open. Now gases can enter and leave the cell.

Oxygen diffuses out of the stomata and into the atmosphere.

Water evaporates out of the stomata and into the atmosphere. This is transpiration.

Carbon dioxide diffuses in through the stomata and then into the cells of the plant.

Nighttime

Most plants close their stomata at night when there is no sunlight for photosynthesis.

When the stomata are closed, gases are unable to enter or leave the cell.

PLANT HORMONES

Plants respond to changes in the environment using molecules called hormones. **Auxin** is a plant growth hormone. Auxin controls growth at the tips of roots and shoots. It has opposite effects in roots and shoots, which affects the way a plant responds to gravity.

Gravitropism

Plants need their roots to grow down so they can absorb water and nutrients from the soil.

If a root is growing sideways under the ground, gravity causes auxin to sink to the lower side of the root. There is more auxin on the underside of the root compared with the upper side of the root.

The way a plant responds to gravity is called **gravitropism**.

Plants need their shoots to grow upward so their leaves can absorb energy from the sun and produce food through photosynthesis.

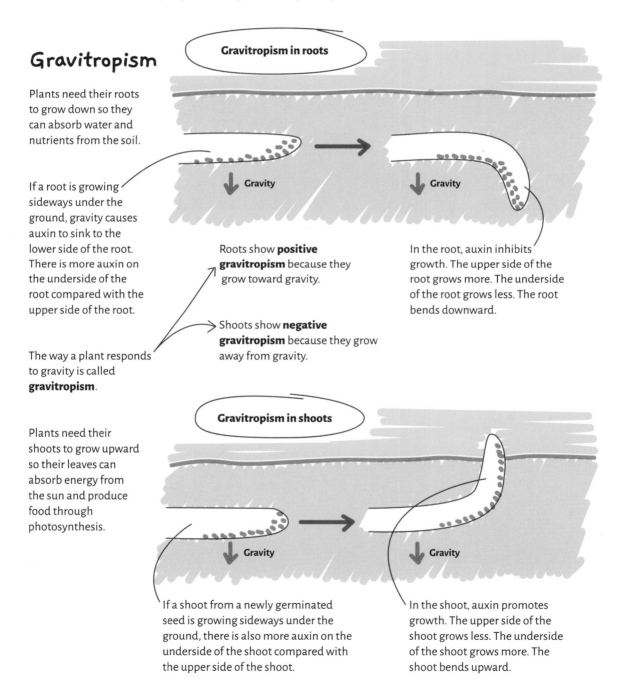

Gravitropism in roots

Gravity

Gravity

Roots show **positive gravitropism** because they grow toward gravity.

In the root, auxin inhibits growth. The upper side of the root grows more. The underside of the root grows less. The root bends downward.

Shoots show **negative gravitropism** because they grow away from gravity.

Gravitropism in shoots

Gravity

Gravity

If a shoot from a newly germinated seed is growing sideways under the ground, there is also more auxin on the underside of the shoot compared with the upper side of the shoot.

In the shoot, auxin promotes growth. The upper side of the shoot grows less. The underside of the shoot grows more. The shoot bends upward.

Phototropism

Plants respond differently to light. This response is called **phototropism**. It is also controlled by the hormone auxin. In shoots, auxin causes cells to grow more.

Artificial uses of auxins:

★ Weed killer: When auxins are sprayed onto the leaves of weeds, the cells divide uncontrollably and the plants die.

★ Rooting powder: Gardeners dip cuttings from plants into auxin powder because it stimulates root growth.

Phototropism in shoots

If the plant is lit from above, auxin is spread evenly throughoutthe shoot tip. This causes the tip to grow straight up.

Gibberellins are another group of plant hormone. They stimulate seed germination, stem growth, and flowering. For example, plant growers use gibberellins to speed up germination and to ensure that all the seeds in a batch germinate at the same time. In the cut flower industry, gibberellins are used to make plants flower all year round.

Ethene is also a plant hormone. It is unusual because it is a gas. It influences plant growth and fruit ripening. For example, bananas are often picked while they are green and unripe. On the way to the supermarket, they are treated with ethene. This causes the fruits to ripen, so they are ready for the customer.

If the plant is lit from one side, auxin diffuses away from the light to the shady side of the stem. This makes the cells on the shady side grow more, so the plant bends toward the light.

PLANT DEFICIENCIES, DISEASES, AND DEFENSES

Plants have complex nutritional requirements. When these requirements are not fulfilled, they don't grow well. Plants can also get diseases from different disease-causing organisms, or pathogens. Plants have evolved various defenses to protect themselves from disease.

Deficiencies

Plants need to acquire various minerals from the soil. They do this by using active transport. Most minerals are needed in small amounts, but others, such as nitrogen, magnesium, and potassium, are needed in larger amounts. Without the right balance of minerals, plants do not grow well. These deficiencies manifest in different ways.

Stunted growth

MINERALS AND MINERAL DEFICIENCIES

Mineral	Used by cells to:	Essential for:	Deficiency
Nitrates	Make proteins	Cell growth	Stunted growth Yellow leaves
Phosphates	Make DNA	Respiration Cell growth	Poor root growth Discolored leaves
Potassium compounds	Help enzyme action	Respiration Photosynthesis	Poor flower and root growth Discolored leaves
Magnesium compounds	Make chlorophyll	Photosynthesis	Yellow leaves

Diseases

Plants can be infected by a variety of pests and pathogens, including bacteria, viruses, and fungi. Diseases cause different symptoms, including abnormal leaves, poor growth, rot, and lumps.

Rose black spot is a fungal disease. Leaves develop spots and fall off. This makes photosynthesis difficult, so the plant doesn't grow well. It is treated by removing the infected leaves and spraying the plant with a fungicide.

Spotty and discolored leaves

Tiny, sap-sucking insects called aphids are a major pest of crop plants, such as cabbages and potatoes. They can cause a lot of damage. Insecticides are used to kill them.

Pests

Malformed stems and leaves

Defenses

Plants have evolved a variety of defense mechanisms to help protect against disease.

PHYSICAL DEFENSES

★ The outer, waxy cuticle provides a barrier to pathogens.

★ Cell walls made from cellulose provide an additional barrier to pathogens.

★ Bark is just a thick layer of dead plant cells. It is another layer of protection.

CHEMICAL DEFENSES

★ Some plants produce antimicrobial chemicals, e.g. the herb thyme contains a compound that kills certain viruses, bacteria, and fungi.

★ Some plants produce poisons that deter animals from eating them, e.g. snowdrops and hyacinths make toxic compounds.

MECHANICAL DEFENSES

★ Some plants have thorns, spines, and prickles that make them hard to eat, e.g. ornamental roses have thorns.

★ Some plants have leaves that droop or curl when an insect lands on them. This makes the insect fall off. The leaves of the mimosa fold up and droop when they are touched.

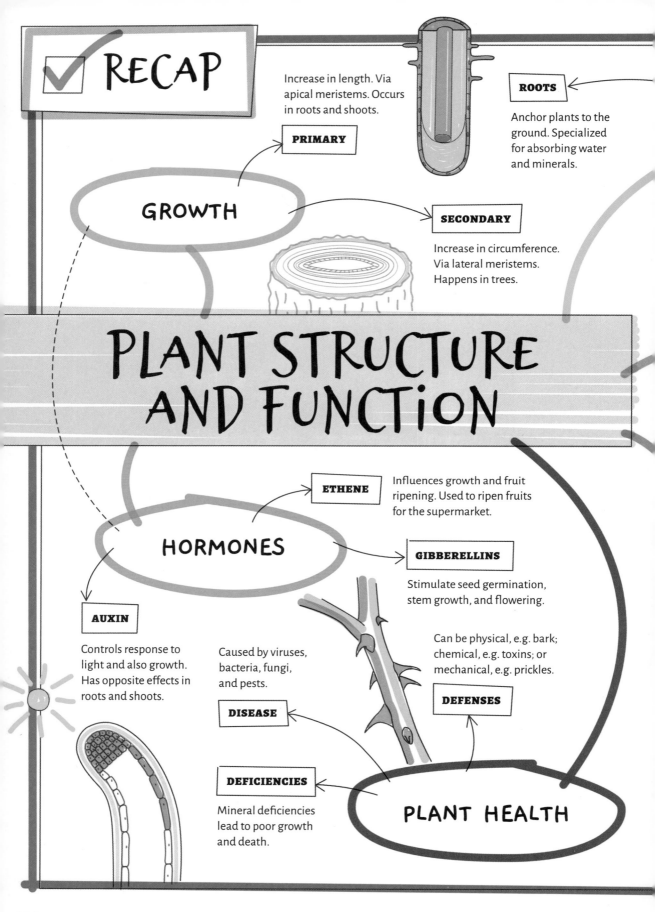

PRIMARY

Increase in length. Via apical meristems. Occurs in roots and shoots.

ROOTS

Anchor plants to the ground. Specialized for absorbing water and minerals.

GROWTH

SECONDARY

Increase in circumference. Via lateral meristems. Happens in trees.

PLANT STRUCTURE AND FUNCTION

ETHENE

Influences growth and fruit ripening. Used to ripen fruits for the supermarket.

HORMONES

GIBBERELLINS

Stimulate seed germination, stem growth, and flowering.

AUXIN

Controls response to light and also growth. Has opposite effects in roots and shoots.

Caused by viruses, bacteria, fungi, and pests.

Can be physical, e.g. bark; chemical, e.g. toxins; or mechanical, e.g. prickles.

DEFENSES

DISEASE

DEFICIENCIES

Mineral deficiencies lead to poor growth and death.

PLANT HEALTH

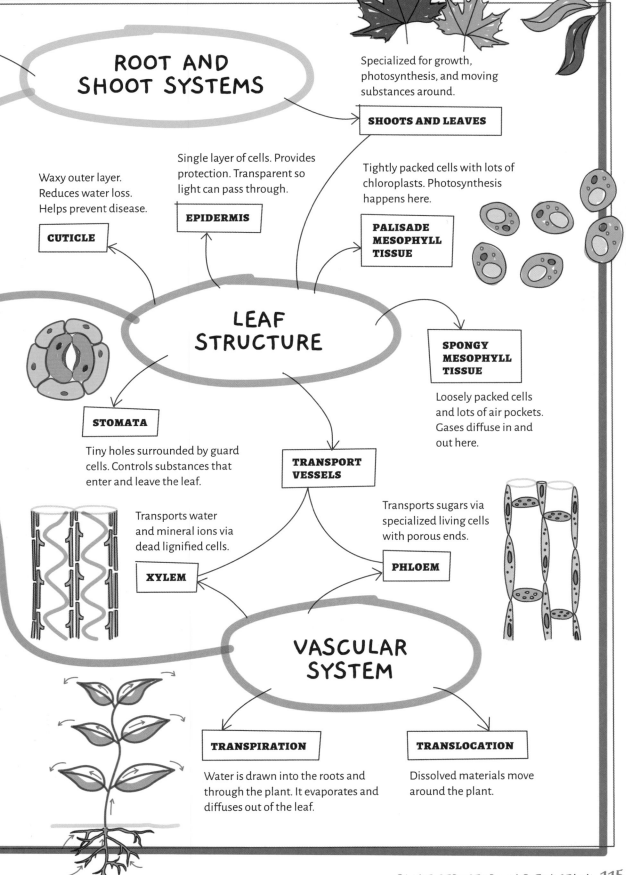

ROOT AND SHOOT SYSTEMS

Specialized for growth, photosynthesis, and moving substances around.

SHOOTS AND LEAVES

Single layer of cells. Provides protection. Transparent so light can pass through.

EPIDERMIS

Waxy outer layer. Reduces water loss. Helps prevent disease.

CUTICLE

Tightly packed cells with lots of chloroplasts. Photosynthesis happens here.

PALISADE MESOPHYLL TISSUE

LEAF STRUCTURE

SPONGY MESOPHYLL TISSUE

Loosely packed cells and lots of air pockets. Gases diffuse in and out here.

STOMATA

Tiny holes surrounded by guard cells. Controls substances that enter and leave the leaf.

TRANSPORT VESSELS

Transports sugars via specialized living cells with porous ends.

Transports water and mineral ions via dead lignified cells.

XYLEM

PHLOEM

VASCULAR SYSTEM

TRANSPIRATION

Water is drawn into the roots and through the plant. It evaporates and diffuses out of the leaf.

TRANSLOCATION

Dissolved materials move around the plant.

CHAPTER 8

HUMAN STRUCTURE AND FUNCTION

In this chapter, you'll learn how the human body maintains a relatively constant internal state despite the environment changing almost constantly. You'll examine the specialized sense organs that help us detect external change. The brain, for example, processes sensory information and coordinates responses using nerves and hormones, while specialized human organ systems provide vital functions, such as breathing and digestion. Similar systems exist in other animals, too, where they work together to help keep organisms alive.

HUMAN ORGAN SYSTEMS

Animal bodies are made up of cells that are organized into tissues and organs, which function together as systems. The human body contains 11 different organ systems.

 **Nervous system**: Responds to environmental stimuli, transmits electrical signals around the body, and organizes the appropriate responses.

 **Skeletal system**: All of the bones and joints in the body, which provide support and protection, enable movement, and produce blood cells.

 **Endocrine system**: A collection of hormone-secreting glands that regulate many functions, including growth, development, and metabolism.

 **Muscular system**: Helps to maintain posture, control movement, and keep joints stable.

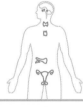 **Digestive system**: Breaks down food into smaller and smaller pieces, so that nutrients can be absorbed into the body.

Lymphatic and immune system: Helps protect the body from disease by mounting an immune response and transporting white blood cells around the body.

 **Circulatory system**: A network involving the heart, blood vessels, and blood, which supplies the body with oxygen and nutrients, transports hormones, and removes waste products.

Urinary system: Removes waste products from the body as urine. Regulates water and salt levels in the blood.

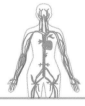 **Respiratory system**: A group of organs, including the lungs, which enable gaseous exchange to occur. Oxygen is transported into the body and carbon dioxide is expelled.

 **Integumentary system**: Includes the skin, hair, and nails; their main function is to act as a barrier and protect the body from the outside world.

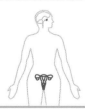 **Female reproductive system**: Produces eggs. Helps to nourish and protect developing embryos.

 **Male reproductive system**: Produces sperm that can then fertilize eggs.

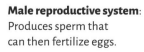

HOMEOSTASIS

In order to function well, organisms must be able to control the conditions inside their body. For example, body temperature and blood glucose levels must be kept within certain limits. **Homeostasis** is the ability of living things and cells to maintain a stable, relatively constant internal environment, even when external conditions are changing.

Homeostasis is controlled automatically. This means you don't need to think about it. Homeostasis is achieved by **automatic control systems**; namely hormones and the nervous system. Automatic control systems contain a sensor, an effector, and a control system.

Homeostasis is maintained using a process called **negative feedback**. For example, if body temperature rises, control centers reduce the temperature. If body temperature falls, control centers increase the temperature.

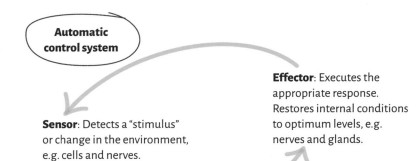

Automatic control system

Sensor: Detects a "stimulus" or change in the environment, e.g. cells and nerves.

Effector: Executes the appropriate response. Restores internal conditions to optimum levels, e.g. nerves and glands.

Control center: Interprets the incoming information and coordinates the appropriate response, e.g. the brain and the pancreas.

Regulating blood glucose

Two hormones, called insulin and glucagon, help regulate glucose levels in the blood. They are both produced by the pancreas. Negative feedback operates to help keep blood glucose levels in the normal range.

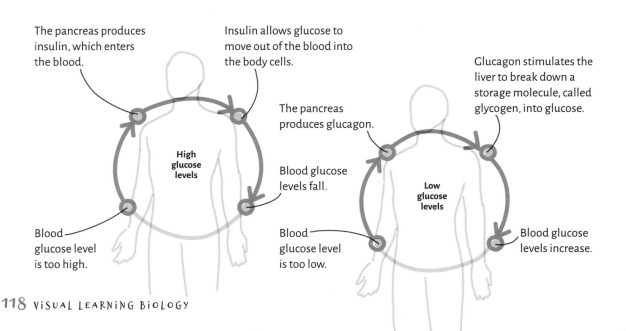

The pancreas produces insulin, which enters the blood.

Insulin allows glucose to move out of the blood into the body cells.

Glucagon stimulates the liver to break down a storage molecule, called glycogen, into glucose.

The pancreas produces glucagon.

High glucose levels

Blood glucose levels fall.

Low glucose levels

Blood glucose level is too high.

Blood glucose level is too low.

Blood glucose levels increase.

Regulating body temperature

The human body functions best at around 37°C (98.6 °F). If a person gets too hot or cold, the change is detected and a number of responses occur.

In the brain, the **hypothalamus** interprets information from the thermoreceptors. It responds by sending nerve impulses to various effectors.

Specialized sensor cells called **thermoreceptors** detect the change in temperature.

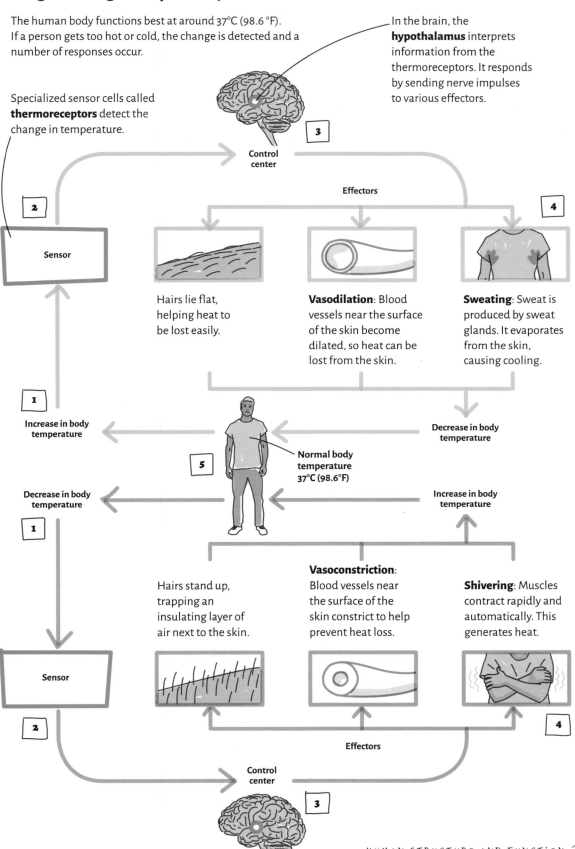

3
Control center

Effectors

2
Sensor

4

Hairs lie flat, helping heat to be lost easily.

Vasodilation: Blood vessels near the surface of the skin become dilated, so heat can be lost from the skin.

Sweating: Sweat is produced by sweat glands. It evaporates from the skin, causing cooling.

1
Increase in body temperature

Decrease in body temperature

5
Normal body temperature 37°C (98.6°F)

Decrease in body temperature

1

Increase in body temperature

Hairs stand up, trapping an insulating layer of air next to the skin.

Vasoconstriction: Blood vessels near the surface of the skin constrict to help prevent heat loss.

Shivering: Muscles contract rapidly and automatically. This generates heat.

2
Sensor

4

Effectors

3
Control center

HUMAN NERVOUS SYSTEM

The **nervous system** helps us react to the world around us, and it plans and coordinates our behavior. It is made of the brain and spinal cord, and all of the many thousands of nerves or **neurons** that connect these organs with the rest of the body. Nerves relay information using electricity and chemicals called neurotransmitters.

Central nervous system (CNS): Contains the brain and spinal cord. This is the body's control center. It receives information from sensor cells called receptors, then coordinates a response via effectors.

Peripheral nervous system (PNS): Contains all of the nerves that connect the CNS to the rest of the body. Nerves in the PNS detect external change and carry out instructions from the CNS.

Synapse: The gap that exists between two adjacent neurons.

Motor neurons: Neurons that carry electrical impulses from the CNS to effectors.

Sensory neurons: Neurons that carry electrical impulses from receptors located around the body to the CNS.

Receptors: Specialized sensor cells that detect environmental change, e.g. touch receptors in the fingers, taste receptors in the mouth, light receptors in the eye.

Effectors: Any of the muscles and glands that respond to electrical impulses, e.g. muscles contract or relax.

An electrical impulse travels along one neuron.

Synapse

When the electrical impulse reaches the synapse, it triggers the release of chemical messengers called **neurotransmitters**. Neurotransmitters, such as dopamine and serotonin, diffuse across the gap and bind to specialized sites on the next neuron.

Now the electrical impulse can continue on its way.

Reflexes

Neurons transmit information rapidly. Sometimes, this process involves conscious thought, but sometimes it happens automatically. The conscious part of the brain is bypassed so a reaction can occur even more quickly. This is called a **reflex**. It is an automatic response. For example, if someone shines a bright light in your eyes, your pupils contract without you having to think about it.

Electrical impulses are transmitted along a sensory neuron toward the CNS.

2

3

Inside the spinal cord, interneurons (relay neurons) bridge the gap between sensory and motor neurons.

4

When the impulse reaches the synapse with the interneuron, neurotransmitters diffuse across the gap and trigger a new nerve impulse, which travels along the interneuron.

A stimulus, such as a hot pan, is detected by specialized pain receptors in the skin.

1

When the impulse reaches the synapse between the interneuron and the motor neuron, neurotransmitters diffuse across the gap and trigger a new nerve impulse.

5

The nerve impulse travels along the motor neuron.

6

The impulse travels to the effector; in this case, a biceps muscle. This causes the muscle to contract. Without thinking, you move your hand away from the hot pan.

7

When an electrical impulse follows a pathway like this, it is called a **reflex arc**.

HUMAN BRAIN

The human brain is the most sophisticated computer in the world. It contains billions of neurons that communicate with one another via trillions of junctions called synapses. It controls everything from thinking, learning, and feeling to seeing, breathing, and moving.

The adult human brain weighs around 1.3 kg (3 lb), which is about the same as a pet rabbit. A fresh brain is soft and squishy, like thick oatmeal. It is protected inside the head by tough membranes called **meninges** and a layer of bone called the **skull**. The human brain is organized into different regions. Each region has a different function.

External parts of the brain

The **frontal lobe** contains the **motor cortex**, which helps control body movements, and the **prefrontal cortex**, which helps with thinking and problem-solving. It also contains **Broca's area**. People with damage to Broca's area have difficulty speaking.

The **parietal lobe** contains the **somatosensory cortex**, which interprets sensations, such as touch, temperature, and pressure. So, the motor cortex lets you physically pick up a cup of coffee, but the somatosensory cortex tells you that the cup is hot.

The **occipital lobe** processes visual information from the eyes.

The **cerebellum** helps coordinate posture, balance, movement, and speech.

The **temporal lobe** processes sound. It contains **Wernicke's area,** which is involved in understanding language. Someone with damage to this area may be able to speak quite normally but not make sense.

The **brain stem** is made up of several smaller structures. It controls involuntary activities.

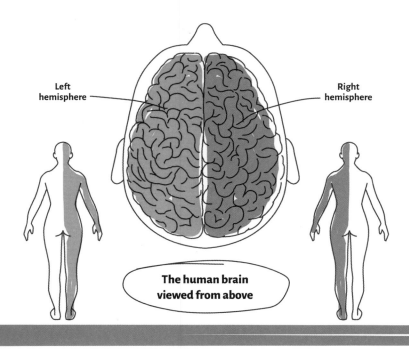

Left hemisphere

Right hemisphere

The **cerebrum** is the gray, wrinkly layer of tissue found on top of the brain. It helps control voluntary actions, such as thinking, speaking, and moving.

The surface of the cerebrum is called the **cortex**. It is divided down the middle into two halves or hemispheres. The **left hemisphere** controls the right side of the body, and the **right hemisphere** controls the left side of the body.

The human brain viewed from above

Internal parts of the brain

The **hypothalamus** links the nervous system to the endocrine system via the pituitary gland. It makes and releases hormones. It controls things like body temperature, hunger, thirst, and blood pressure.

The **thalamus** processes information going to and from the spinal cord.

The **cerebellum** helps to control balance, coordination, and muscular activity.

The **hippocampus** is involved in learning and memory.

The **pituitary gland** helps maintain homeostasis by secreting hormones.

The **medulla** is part of the brain stem. It controls unconscious activities, such as breathing, heart rate, and blood pressure. Specialized cells here also control sneezing and vomiting.

SENSE ORGANS

Our bodies contain various different sense organs, such as the eyes, ears, and tongue, which help us to detect change in the environment. Sense organs relay information to the central nervous system, which enables us to process and respond to these changes.

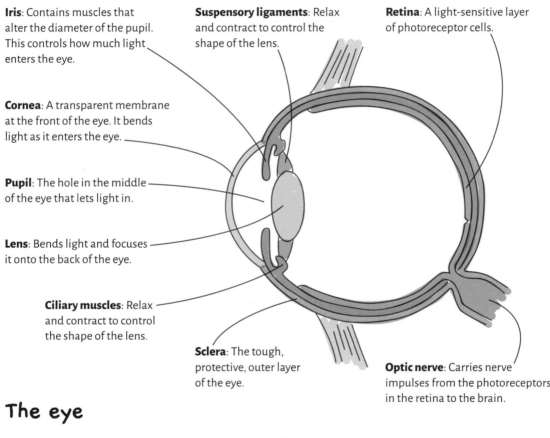

Iris: Contains muscles that alter the diameter of the pupil. This controls how much light enters the eye.

Cornea: A transparent membrane at the front of the eye. It bends light as it enters the eye.

Pupil: The hole in the middle of the eye that lets light in.

Lens: Bends light and focuses it onto the back of the eye.

Ciliary muscles: Relax and contract to control the shape of the lens.

Suspensory ligaments: Relax and contract to control the shape of the lens.

Sclera: The tough, protective, outer layer of the eye.

Retina: A light-sensitive layer of photoreceptor cells.

Optic nerve: Carries nerve impulses from the photoreceptors in the retina to the brain.

The eye

The eye is an important sense organ. It can detect complex stimuli, such as contrast, color, and movement. The eye may seem to be a discrete unit, but it is actually an extension of the central nervous system. This is because the optic nerve, which runs out of the back of the eye, feeds directly into the brain.

Rods: A type of cell found mainly in the periphery of the retina. They are used for peripheral vision, and they work well in dim light.

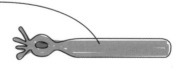

Cones: A type of cell found in the retina. Cones produce color vision. They work best in bright light. The human eye has three different types of cones: red, green, and blue. They are sensitive to different parts of the visible light spectrum.

THE IRIS REFLEX

Very bright light can damage the eye. When photoreceptors in the retina detect very bright light, it triggers a reflex that makes the pupil smaller.

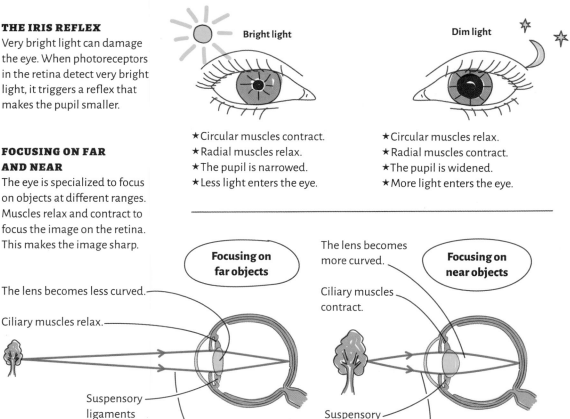

Bright light

★ Circular muscles contract.
★ Radial muscles relax.
★ The pupil is narrowed.
★ Less light enters the eye.

Dim light

★ Circular muscles relax.
★ Radial muscles contract.
★ The pupil is widened.
★ More light enters the eye.

FOCUSING ON FAR AND NEAR

The eye is specialized to focus on objects at different ranges. Muscles relax and contract to focus the image on the retina. This makes the image sharp.

Focusing on far objects

The lens becomes less curved.

Ciliary muscles relax.

Suspensory ligaments contract.

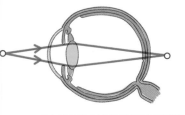

Light entering the eye is bent less.

The lens becomes more curved.

Focusing on near objects

Ciliary muscles contract.

Suspensory ligaments relax.

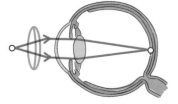

Light entering the eye is bent more.

FAR- AND NEAR-SIGHTEDNESS

Far- and near-sightedness can be corrected using lenses.

Far-sighted (hyperopic)

Far-sighted people struggle to focus on near objects.

Images of close objects are focused behind the retina. This makes them look blurry.

This can be corrected with convex lenses, which focus the light rays on the retina.

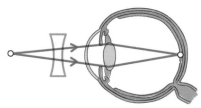

Near-sighted (myopic)

Near-sighted people struggle to focus on distant objects.

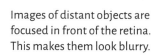

Images of distant objects are focused in front of the retina. This makes them look blurry.

This can be corrected with concave lenses, which focus the light rays on the retina.

The ear

The ear is specialized to receive and focus sound waves, convert them into electrical signals, and relay the signals to the brain.

Outer ear or pinna: The visible part of the ear. Shaped like a funnel to gather sound waves and send them to the inner ear.

Hammer (malleus): One of three tiny bones in the middle ear. Vibrates in response to sound waves.

Anvil (incus): One of three tiny bones in the middle ear. It vibrates when the hammer moves.

Stirrup (stapes): One of three tiny bones in the middle ear. It vibrates when the anvil moves.

Semicircular canals: Three tiny, fluid-filled tubes that are involved in balance rather than hearing.

Cochlea: A spiral cavity filled with liquid that moves when the stirrup vibrates. This triggers electrical impulses.

Middle ear

Auditory nerve: Transports electrical impulses from the cochlea to the brain, where they are interpreted as sound.

Ear drum (tympanum): A thin, transparent membrane that vibrates in response to sound waves.

Ear canal: The tube that connects the outer and inner ear. Sound waves travel along this tube. It is lined with cells that produce ear wax. Ear wax helps to protect and moisturize the skin of the ear canal, and provides some protection against bacteria.

Eustachian tube: A tube that connects the middle ear to the throat and nasal cavity. It controls the pressure inside the middle ear.

Outer ear

Inner ear

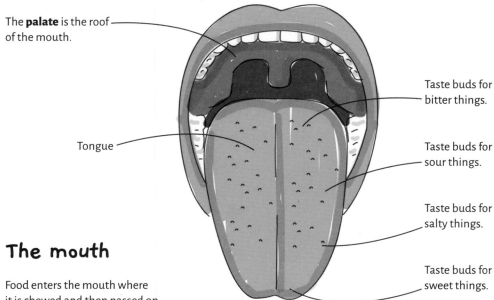

The **palate** is the roof of the mouth.

Taste buds for bitter things.

Taste buds for sour things.

Tongue

Taste buds for salty things.

Taste buds for sweet things.

The mouth

Food enters the mouth where it is chewed and then passed on to the rest of the digestive system. Before it is swallowed, it is tasted. The **tongue** contains a variety of specialized taste cells, which help us to enjoy our food.

The tongue is a muscular organ in the mouth, used for chewing and swallowing food, and for making sounds. It is kept moist by saliva, and it is supplied with many nerves and blood vessels. The upper surface of the tongue is covered in tiny bumps called **papillae**. Some are called **mechanical papillae**, because they help us feel the texture of food. Others are called **taste papillae**, because they help us to taste food.

Other sense organs

There are five classic senses—sight, hearing, taste, smell, and touch—but scientists think there could actually be as many as 50 different senses.

Although there are obvious sense organs, such as the eyes, ears, and nose, there are many specialized sensory cells throughout the body.

These are equipped to recognize other sensations, in addition to the classic five, for example:

★ **Hunger** is the sense of needing to eat.

★ **Thermoception** is the ability to perceive different temperatures.

★ **Nociception** is the perception of pain.

★ **Equilibrioception** is the ability to perceive balance.

★ **Proprioception** is the ability to know where your different body parts are, e.g. being able to touch your nose while your eyes are closed.

ENDOCRINE SYSTEM

The **endocrine system** helps the body to respond to changes in the environment. It is made of various glands. The glands release chemical molecules called **hormones** into the bloodstream. These then travel around the body and have an effect on specific organs.

Thyroid gland: Found in the neck where it produces thyroid hormones, such as thyroxine. Thyroid hormones influence metabolism. In children, they also influence growth and development.

Pituitary gland: A pea-sized gland found at the base of the brain. It secretes many different hormones, e.g. prolactin, which stimulates milk production in the breasts, and growth hormone, which regulates growth and metabolism. It is often called the "master gland" because it controls the activity of other glands, e.g. pituitary hormones act on the adrenal glands, thyroid glands, ovaries, and testes, which then produce other hormones.

Adrenal glands: Sit just above the kidneys and produce a variety of hormones, e.g. adrenaline, which prepares the body to quickly respond to threats ("fight or flight"), and cortisol, which influences blood sugar levels and metabolism.

Pancreas: As well as making digestive juices, the pancreas secretes hormones, such as insulin and glucagon, which help control blood glucose levels.

Ovaries (females only): Produce estrogen which influences the development of breasts, pubic hair, and other secondary sexual characteristics, as well as the menstrual cycle and fertility.

Testes (males only): Produce testosterone which influences the development of facial hair, the Adam's apple, and other secondary sexual characteristics, as well as the production of sperm.

The fight or flight response

Most hormones have a prolonged effect, but sometimes the effect can be immediate. The fight or flight response is a good example.

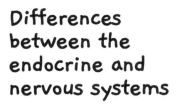

Threat: A dangerous or stressful situation is perceived.

Brain: Processes the signal.

Adrenal glands: Secrete adrenaline and cortisol into the bloodstream.

Physical effects: The hormones prepare the body, either to defend itself (fight) or run away (flight).

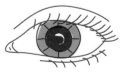

Differences between the endocrine and nervous systems

The endocrine system influences specific tissues via hormones that circulate in the blood. The hormones act slowly over a long time, until they are broken down, and the responses tend to influence large areas of the body. In contrast, the nervous system influences muscles and glands via nerves. The nerve impulses act rapidly for a short time only, and responses are localized to specific areas of the body.

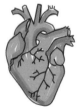

Heart rate and blood pressure increase. More blood is pumped into vital organs, such as the brain and muscles, preparing the body for action.

Blood is redirected to vital organs so extremities, like the hands and feet, may feel cold and clammy.

Digestion is slowed, so the body can focus on the fight or flight reaction.

The pupils dilate so more light can enter the eye, and vision improves.

The pain response is blunted, so people can sometimes ignore injuries.

In really stressful circumstances, some people lose control of their bladder.

HUMAN DIGESTIVE SYSTEM

Mouth

The mouth is where digestion begins. Chewing breaks down food into smaller pieces. Taste buds detect the chemical makeup of your food, prompting the cells of the digestive system to release food-digesting enzymes. An enzyme found in saliva—salivary amylase—starts to chemically digest any carbohydrates that are present.

Esophagus

The esophagus is a muscular tube. It connects the mouth with the stomach. Waves of muscular contractions squeeze the food along the esophagus into the stomach. This is called **peristalsis**. The same process also helps to move food through the stomach and intestines.

Stomach

The stomach helps to store and digest food. Up to 2 liters (4 pints) of food and fluid can be stored inside this stretchy organ. Gastric juice, secreted by glands in the stomach lining, turns small pieces of food into a soupy suspension. Gastric juice contains hydrochloric acid, which makes proteins unravel, and enzymes called proteases, which break these exposed proteins down into smaller molecules.

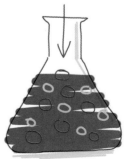

The adult **human digestive system** can be up to 9 meters (30 feet) long. It contains lots of different organs. As food journeys through the digestive system, it is broken down into smaller pieces. This is done mechanically, by chewing, and chemically, by various enzymes that are released into the digestive tract. Once the food is broken down, the body absorbs useful molecules. This provides us with energy. Any non-digestible food is eliminated.

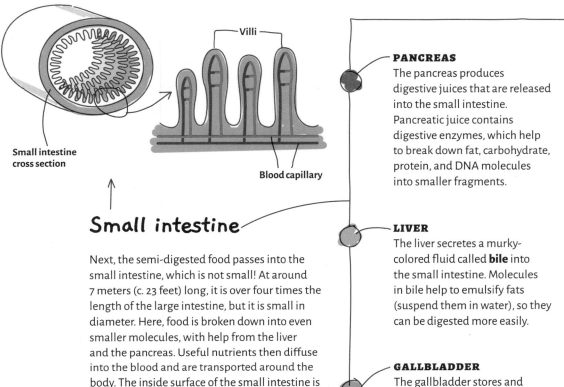

Villi

Small intestine
cross section

Blood capillary

Small intestine

Next, the semi-digested food passes into the
small intestine, which is not small! At around
7 meters (c. 23 feet) long, it is over four times the
length of the large intestine, but it is small in
diameter. Here, food is broken down into even
smaller molecules, with help from the liver
and the pancreas. Useful nutrients then diffuse
into the blood and are transported around the
body. The inside surface of the small intestine is
covered in tiny folds called villi. These have even
tinier folds on them, called microvilli. These
increase the surface area for digestion.

PANCREAS

The pancreas produces
digestive juices that are released
into the small intestine.
Pancreatic juice contains
digestive enzymes, which help
to break down fat, carbohydrate,
protein, and DNA molecules
into smaller fragments.

LIVER

The liver secretes a murky-
colored fluid called **bile** into
the small intestine. Molecules
in bile help to emulsify fats
(suspend them in water), so they
can be digested more easily.

GALLBLADDER

The gallbladder stores and
concentrates bile before it is
released into the small intestine.

APPENDIX

The appendix is a
small, pouch-like
sac of tissue. It is
thought to play
a role in the
immune system.

Large intestine

Undigested food is squeezed into the
large intestine, where water and
some electrolytes are then absorbed
into the bloodstream.

Rectum

After the large intestine, any
undigested material passes into
the rectum. This leftover material
forms the feces that we produce.
The rectum stores feces.

Anus

The anus is the last part of the
digestive system. It is the external
opening of the rectum. Feces are
expelled via the anus at an
opportune moment.

CIRCULATORY AND RESPIRATORY SYSTEMS

The circulatory and respiratory systems work together to pump blood around the body, expel carbon dioxide, and provide oxygen for respiration. The circulatory system contains the heart, blood vessels, and blood. The main organs of the respiratory system are the lungs.

The heart

The heart is one of the body's vital organs. It contains four chambers: two atria and two ventricles. The walls of the heart are made of muscles, which contract to pump blood around the body. It also contains valves, which ensure that blood flows in the right direction.

How the heart works:

1 Oxygenated blood enters the heart via the pulmonary vein. Deoxygenated blood enters the heart via the vena cava.

2 The atria contract, pushing the blood into the ventricles.

3 The ventricles contract, pushing oxygenated blood into the aorta and deoxygenated blood into the pulmonary artery.

4 Blood flows to the organs via arteries and returns via veins.

5 The atria fill up with blood and the cycle starts again.

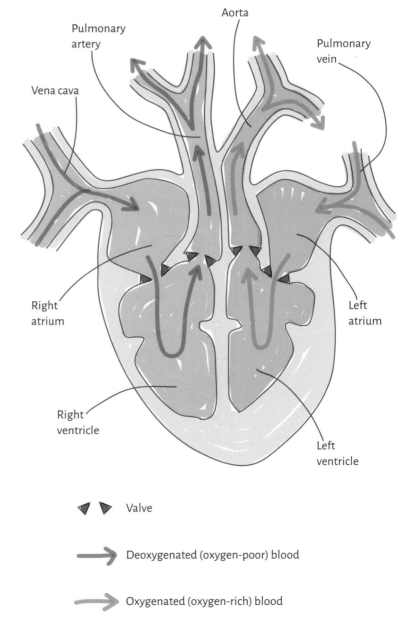

Aorta

Pulmonary artery

Pulmonary vein

Vena cava

Right atrium

Left atrium

Right ventricle

Left ventricle

◀ ▶ Valve

⟶ Deoxygenated (oxygen-poor) blood

⟶ Oxygenated (oxygen-rich) blood

The double circulatory system

The circulatory system is made up of two circuits joined together.

The wall of the left ventricle is thicker than the wall of the right ventricle, because it has to pump blood much further, all around the systemic circuit.

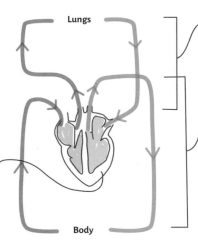

Lungs

Body

The pulmonary circuit: The right ventricle pumps deoxygenated blood to the lungs where it acquires oxygen. Oxygenated blood then returns to the heart.

The systemic circuit: The left ventricle pumps oxygenated blood around the body. As cells and organs use up oxygen, the blood becomes deoxygenated. It then returns to the heart so it can be pumped to the lungs.

The lungs

The lungs are specialized to allow gas exchange to occur. The circulatory system delivers deoxygenated blood to the lungs, where it is oxygenated and then returned to the body via the heart.

The respiratory system

Trachea (windpipe): A large tube reinforced with rings of cartilage. Connects the mouth to the bronchi.

Pleural membrane: A moist membrane that surrounds the lungs and keeps them airtight.

Bronchi: There are two bronchi. They are the major airway passages that connect the trachea to the lungs.

Bronchioles: The bronchi split into smaller and smaller tubes called bronchioles.

Alveoli: The tips of the bronchioles end in these tiny air sacs. This is where gas exchange occurs. Oxygen diffuses into neighboring blood vessels, and carbon dioxide diffuses out of neighboring blood vessels.

Intercostal muscles: Muscles that contract and relax to move the ribcage and help breathing.

Diaphragm: A large sheet of muscle that contracts and relaxes to aid breathing.

Blood vessels

Blood moves around the body in blood vessels. There are three types of blood vessels: capillaries, veins, and arteries.

Capillaries
★ Are a huge network of tiny blood vessels.
★ Have thin walls, just one cell thick. This allows gaseous exchange to occur.

Veins
★ Carry deoxygenated blood (except for the pulmonary vein) at low pressure from the body to the heart.
★ Have thin walls that contain valves to keep blood flowing in the right direction.
★ Have wide channels so that blood can flow easily.

Arteries
★ Carry oxygenated blood (except for the pulmonary artery) at high pressure from the heart to the body.
★ Have thick, muscular walls.
★ Have narrow channels that help maintain high pressure.

SKELETAL AND MUSCULAR SYSTEMS

The skeletal system offers support and helps to protect vital organs, such as the brain and the heart. It also makes blood cells. It is assisted by the muscular system, which also helps to control movement and maintain posture.

The human skeleton

The **human skeleton** consists of more than 200 bones. Each bone is a living tissue with its own blood supply. Bones contain calcium and other minerals that help to make them strong and slightly flexible.

Humerus: Connects the arm to the shoulder. Helps the arm to move.

Radius and ulna: The two bones of the forearm. Connect the arm to the hand. Help the arm to move.

Femur (thighbone): The longest bone in the body. Supports the weight of the body and enables the leg to move.

Tibia (shinbone) and **fibula**: The two lower leg bones. They help the leg and feet to move.

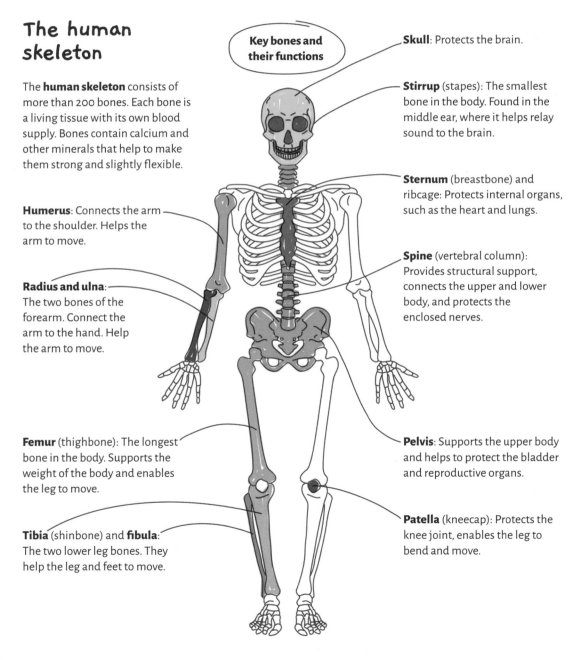

Key bones and their functions

Skull: Protects the brain.

Stirrup (stapes): The smallest bone in the body. Found in the middle ear, where it helps relay sound to the brain.

Sternum (breastbone) and ribcage: Protects internal organs, such as the heart and lungs.

Spine (vertebral column): Provides structural support, connects the upper and lower body, and protects the enclosed nerves.

Pelvis: Supports the upper body and helps to protect the bladder and reproductive organs.

Patella (kneecap): Protects the knee joint, enables the leg to bend and move.

Joints

Joints link bones together and help the skeleton to move. Hinge joints, such as the knee and elbow, help bones to move back and forth. Ball and socket joints, such as the shoulder and hip, enable bones to rotate in multiple directions.

A synovial joint

Synovial fluid: A liquid that reduces friction between the bones.

Synovial membrane: Specialized connective tissue that lines the inside of the joint.

Cartilage: A tough, smooth substance that covers the ends of the bone. It protects against wear and tear.

Ligaments: Tough, connective tissue that connects bones to bones, and holds the joint together.

Muscles

Joints cannot work without muscles. Muscles are a type of specialized tissue. They are attached to bones by **tendons**, which are a type of connective tissue. Muscles move bones by contracting. They often work in pairs.

The elbow joint is controlled by two muscles that work **antagonistically**. This means when the biceps contract, the triceps relax, and vice versa.

Blood cells

Big bones contain soft tissue called **bone marrow**. Bone marrow makes blood cells. There are different types of blood cells:

Red blood cells: Specialized cells that transport oxygen around the body. They are small and flexible, lack nuclei, and look like flattened discs.

White blood cells: A variety of specialized cells that form part of the immune system, and help to fight disease.

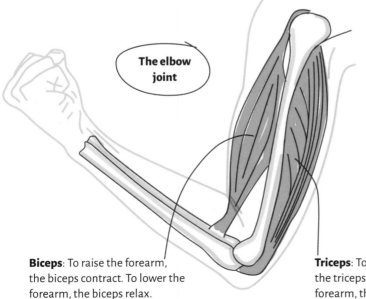

The elbow joint

Biceps: To raise the forearm, the biceps contract. To lower the forearm, the biceps relax.

Triceps: To raise the forearm, the triceps relax. To lower the forearm, the triceps contract.

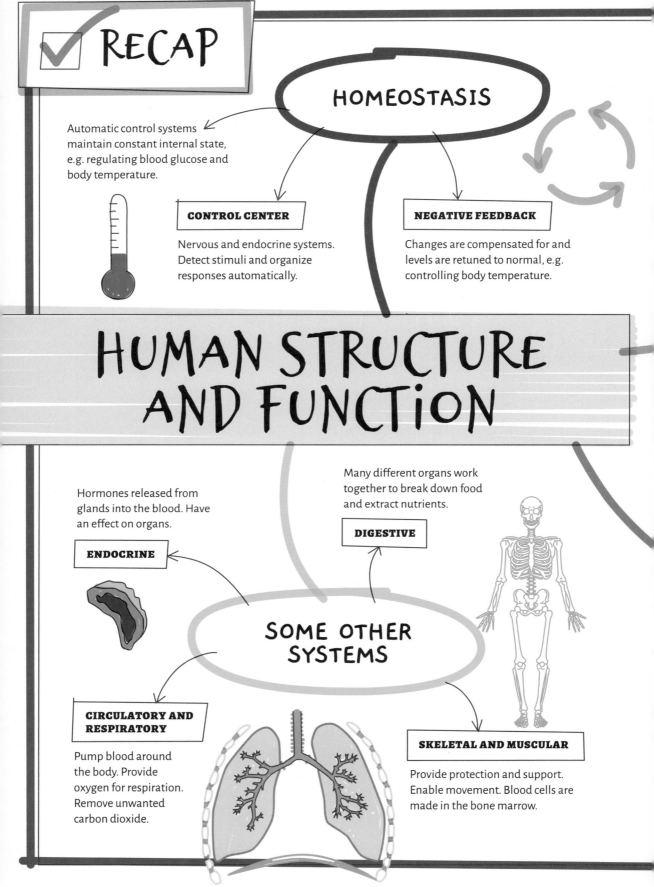

HOMEOSTASIS

Automatic control systems maintain constant internal state, e.g. regulating blood glucose and body temperature.

CONTROL CENTER

Nervous and endocrine systems. Detect stimuli and organize responses automatically.

NEGATIVE FEEDBACK

Changes are compensated for and levels are retuned to normal, e.g. controlling body temperature.

HUMAN STRUCTURE AND FUNCTION

Hormones released from glands into the blood. Have an effect on organs.

ENDOCRINE

Many different organs work together to break down food and extract nutrients.

DIGESTIVE

SOME OTHER SYSTEMS

CIRCULATORY AND RESPIRATORY

Pump blood around the body. Provide oxygen for respiration. Remove unwanted carbon dioxide.

SKELETAL AND MUSCULAR

Provide protection and support. Enable movement. Blood cells are made in the bone marrow.

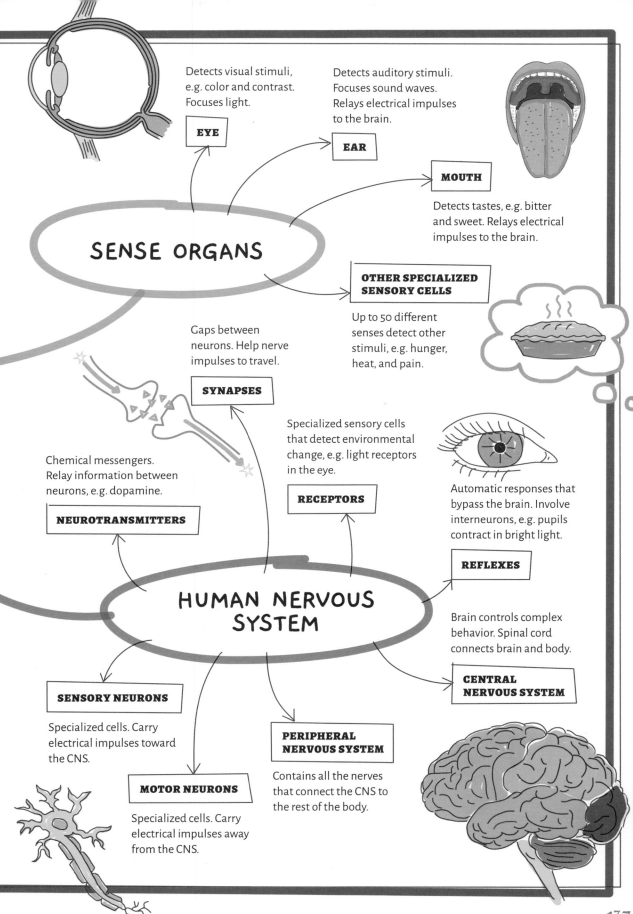

Detects visual stimuli, e.g. color and contrast. Focuses light.

EYE

Detects auditory stimuli. Focuses sound waves. Relays electrical impulses to the brain.

EAR

MOUTH

Detects tastes, e.g. bitter and sweet. Relays electrical impulses to the brain.

SENSE ORGANS

OTHER SPECIALIZED SENSORY CELLS

Up to 50 different senses detect other stimuli, e.g. hunger, heat, and pain.

Gaps between neurons. Help nerve impulses to travel.

SYNAPSES

Specialized sensory cells that detect environmental change, e.g. light receptors in the eye.

RECEPTORS

Chemical messengers. Relay information between neurons, e.g. dopamine.

NEUROTRANSMITTERS

Automatic responses that bypass the brain. Involve interneurons, e.g. pupils contract in bright light.

REFLEXES

HUMAN NERVOUS SYSTEM

Brain controls complex behavior. Spinal cord connects brain and body.

CENTRAL NERVOUS SYSTEM

SENSORY NEURONS

Specialized cells. Carry electrical impulses toward the CNS.

PERIPHERAL NERVOUS SYSTEM

Contains all the nerves that connect the CNS to the rest of the body.

MOTOR NEURONS

Specialized cells. Carry electrical impulses away from the CNS.

CHAPTER 9

HUMAN HEALTH AND DISEASE

People are at their best when they are healthy and happy. You can stay healthy by making positive lifestyle choices, such as exercising and eating well. When diseases do occur, they can be caused by many things, including infectious microorganisms, faulty genes, and environmental factors. Global health inequalities exist, so people in some parts of the world experience more ill health than others. Fortunately, scientists are always developing new medicines and treatments to help prevent, treat, and cure diseases. In this chapter, you'll find out more about human health and disease.

HEALTH INEQUALITIES

A round the world, people experience avoidable, systemic-based differences in their health care. Some groups of people are more likely to experience disease than others due to their circumstances. **Health inequalities** reflect differences in geography, education, and other factors.

Life expectancy is a measure of expected life span. Life expectancy at birth is a good indicator of health later in life. Life expectancy varies by 34 years between countries.

A: High-income countries, e.g. life expectancy in Japan is 84 years.

B: Low-income countries, e.g. life expectancy in Sierra Leone is 50 years.

Darker colors = higher burden of disease. People here have lower life expectancies.

Lighter colors = lower burden of disease. People here have higher life expectancies.

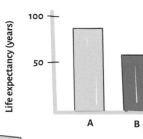

Life expectancy (years)

Health inequalities occur within countries too. In the USA, for example, African Americans account for 13% of the population but experience almost half of all new HIV infections.

Sub-Saharan Africa experiences the highest burden of disease. This is due to poverty and a lack of basic resources.

Geographical and social factors interact. Every day, 16,000 children die before their fifth birthday. These deaths are mostly preventable, and 14 times more likely if the child lives in sub-Saharan Africa. In addition, children from poor and rural households are more likely to die than children from richer, less rural households. Deaths occur from a lack of basic resources, such as food and clean water, and from diseases, such as malaria and measles.

COMMUNICABLE DISEASES

Diseases that spread from person to person or between animals and people are called communicable or infectious **diseases**. They are caused by disease-causing microorganisms called **pathogens**. Some communicable diseases can be relatively mild, such as conjunctivitis. Others can be more serious, for example malaria.

Communicable diseases can be spread in a variety of communicable or infectious ways.

Direct contact with an infected person, e.g. chicken pox spreads if there is close contact between an infected and a susceptible person. HIV can spread via sexual intercourse.

Indirect contact with a contaminated substance or item, e.g. HIV can also be spread when drug users with HIV share their needles.

Airborne transmission, e.g. when someone with a cold sneezes, thousands of tiny droplets are released, full of viral particles, into the air. These can be inhaled by other people who then become infected.

Via water, e.g. the bacterium that causes cholera is spread by contaminated water.

Via vectors: Living organisms can transmit pathogens to humans, e.g. blood-sucking insects, such as mosquitoes, can spread malaria.

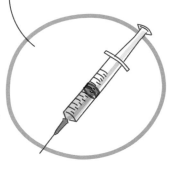

Causes of communicable disease

There are four main types of pathogens that cause communicable diseases: bacteria, fungi, protists, and viruses.

BACTERIA

Our bodies contain trillions of bacteria. Most are harmless but some cause disease. For example, **tuberculosis** (TB) is an airborne bacterial disease that affects the lungs. It occurs in every part of the world and kills more than a million people every year.

Antibiotics slow down and stop the growth of bacteria. Some target specific bacteria. Others have a wider range. Since their discovery, in the 1920s, antibiotics have saved millions of lives.

TB can be cured with antibiotics, but now the bacterium is becoming resistant to some of these drugs. **Antibiotic resistance** is making a growing number of infections, such as pneumonia and gonorrhea, harder to treat. Now antibiotic resistance is one of the biggest threats to global health.

FUNGI

Around 300 types of fungi cause disease in humans. Anti-fungal treatments can help to control many of these. For example, aspergillosis is a fungal infection that affects the respiratory system. Symptoms can be mild to severe, and it is often treated with antifungal drugs.

PROTISTS

Protists cause many different diseases, some of which can be deadly. Often the protist is transferred to the human host by a vector. Malaria is a protist disease. It is caused by *Plasmodium* parasites that are spread to people through mosquito bites. Every year, more than 200 million new cases are reported. People with malaria experience fever, headache, and chills. Left untreated, symptoms worsen. Malaria claims about 400,000 lives every year.

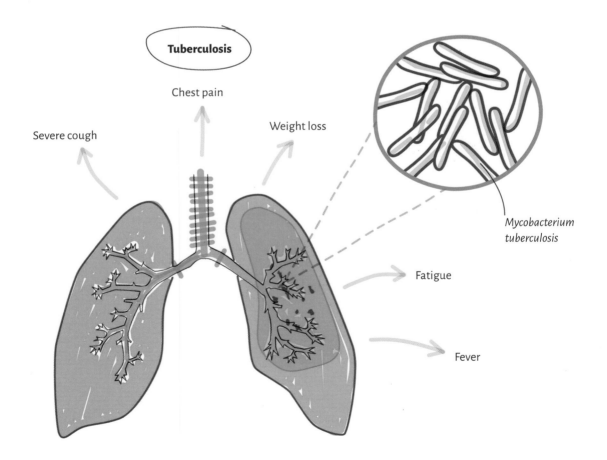

Tuberculosis

Chest pain

Severe cough

Weight loss

Mycobacterium tuberculosis

Fatigue

Fever

VIRUSES

Viruses infect living cells. Viral diseases cannot be treated with antibiotics, and there are very few successful antiviral drugs. For example, coronaviruses are a family of disease-causing viruses that cause respiratory infections. Some of these, such as the common cold, are mild. Others, such as COVID-19, can be fatal.

In 2020, a coronavirus called SARS-CoV-2 triggered a global pandemic. SARS-CoV-2 causes the disease COVID-19.

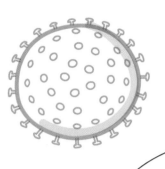

Incubation period: SARS-CoV-2 invades the body. Cells in the throat, airways, and lungs churn out copies of the virus, which infect more cells. The patient has no symptoms at first, but can spread the disease to others. Some people may carry the virus but never develop symptoms.

Mild disease: Most infected people experience mild symptoms, such as fever, dry cough, and a loss of taste or smell. The body produces molecules called cytokines, which spur the immune system into action. Most people recover within a week or two.

People with COVID-19 can experience a wide range of symptoms that affect many different organs. Scientists think this could be because of the way the virus targets cells in the body.

The virus latches on to receptor proteins, known as ACE-2, found on the surface of human cells. When the virus first invades the body, it binds to ACE-2 receptors in the nose, throat, and lungs. Then, as it replicates and spreads, it attaches to ACE-2 receptors in other organs, such as the intestines, heart, and kidneys. It's thought that this leads to symptoms that affect other parts of the body.

Serious disease: Some people develop dangerously high levels of inflammation. In the lungs, this can cause pneumonia. These people may need a ventilator to help them breathe. Sometimes blood clots develop, and the patient needs blood-thinning medication. Although many people recover, some are left with long-term damage to organs, such as the lungs, heart, kidneys, and brain.

Critical disease: This affects a small but significant percentage of cases. The body struggles to receive enough oxygen and organs start to fail. In just the first 6 months of the pandemic, more than 600,000 people around the world died from COVID-19.

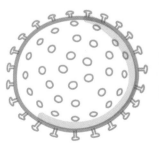

ACE-2 receptor

Preventing the spread of communicable diseases

There are many ways to curb the transmission of communicable diseases.

Social distancing: During the COVID-19 pandemic of 2020, people were asked to reduce contact with others. Schools were closed, and people stayed at home, and kept their distance when in public.

Quarantine: Sometimes infected people are put into isolation so they cannot infect others.

Hygiene: Some infectious diseases are acquired when people touch their face with contaminated hands. Regular hand washing and good hygiene can help prevent this.

Vector control: Malaria is spread by mosquitoes, which can be killed with insecticides. Bed nets protect people from being bitten, and anti-malarial drugs also help to control the disease.

Contraception: Condoms can stop the spread of bodily fluids and sexually transmitted diseases, such as HIV and chlamydia.

Vaccines: Vaccines train the immune system to recognize and destroy pathogens. Vaccinating people against communicable diseases means they will not become infected or spread the disease.

Once enough people are vaccinated, opportunities for disease outbreaks become so low that even people who aren't immunized benefit. This is called **herd immunity**. The pathogen struggles to find enough hosts and eventually dies out.

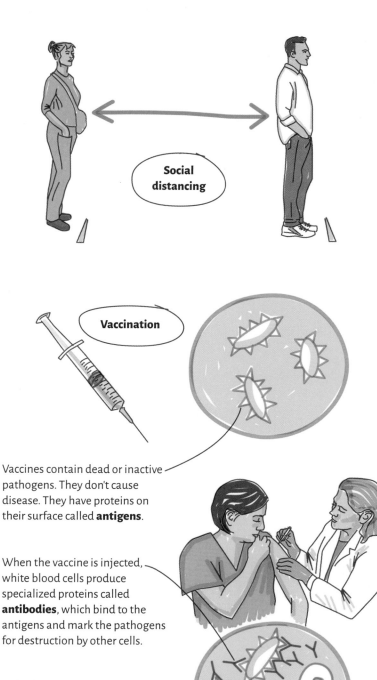

Social distancing

Vaccination

Vaccines contain dead or inactive pathogens. They don't cause disease. They have proteins on their surface called **antigens**.

When the vaccine is injected, white blood cells produce specialized proteins called **antibodies**, which bind to the antigens and mark the pathogens for destruction by other cells.

If the disease is encountered, the immune system is now primed and ready to go. The antigens are immediately recognized, and the pathogens are killed before they can spread and cause sickness.

Antibody   White blood cell

NONCOMMUNICABLE DISEASES

Diseases that are not infectious are called **noncommunicable**. They cannot be spread from person to person. Noncommunicable diseases, such as Alzheimer's disease, diabetes, and cancer, kill 41 million people every year. This accounts for 70% of all deaths globally. They affect people of all ages and all countries, but 85% occur in low- and middle-income countries.

The top four noncommunicable diseases account for more than 80% of all premature deaths caused by these diseases.

Noncommunicable diseases are caused by many different factors. Often multiple factors interact to produce complex causes that are hard to unravel.

Causes of noncommunicable diseases

Faulty genes can cause diseases, such as sickle cell anemia, which occurs when a faulty gene is inherited. Abnormally shaped red blood cells struggle to carry oxygen around the body, leading to anemia and shortness of breath.

DNA damage can occur during the life span instead of being inherited. For example, skin cancer happens when ultraviolet (UV) rays damage the DNA inside skin cells. Cells can repair some acquired DNA damage, but this process is not always perfect.

Vitamin and mineral deficiencies can cause diseases, such as rickets, which is a bone disorder that can occur when there is a lack of vitamin D. Scurvy is caused by a lack of vitamin C.

Environmental and lifestyle factors can cause diseases, such as lung cancer, which can be caused by smoking, and liver disease, which can be caused by drinking too much alcohol.

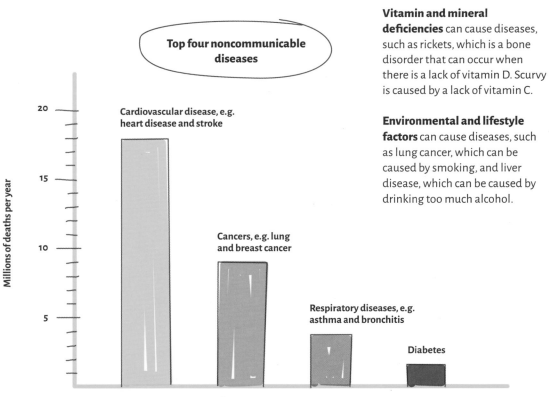

Top four noncommunicable diseases

Millions of deaths per year

20

15

10

5

Cardiovascular disease, e.g. heart disease and stroke

Cancers, e.g. lung and breast cancer

Respiratory diseases, e.g. asthma and bronchitis

Diabetes

Diseases

Cardiovascular diseases

Diseases that affect the heart and blood vessels are called **cardiovascular diseases.** They are the leading cause of death globally. For example, **coronary heart disease** occurs when the coronary arteries that supply blood to the heart become narrowed. This can cause a heart attack.

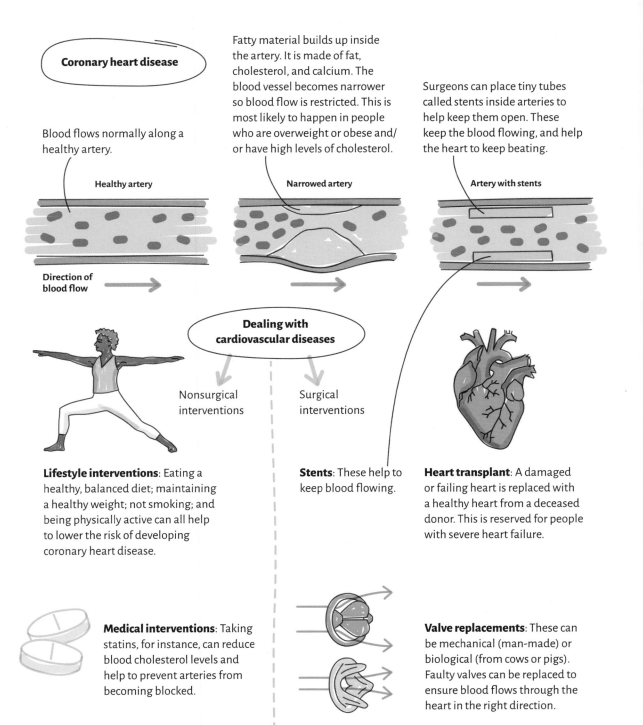

Coronary heart disease

Blood flows normally along a healthy artery.

Healthy artery

Direction of blood flow

Fatty material builds up inside the artery. It is made of fat, cholesterol, and calcium. The blood vessel becomes narrower so blood flow is restricted. This is most likely to happen in people who are overweight or obese and/ or have high levels of cholesterol.

Narrowed artery

Surgeons can place tiny tubes called stents inside arteries to help keep them open. These keep the blood flowing, and help the heart to keep beating.

Artery with stents

Dealing with cardiovascular diseases

Nonsurgical interventions

Surgical interventions

Lifestyle interventions: Eating a healthy, balanced diet; maintaining a healthy weight; not smoking; and being physically active can all help to lower the risk of developing coronary heart disease.

Stents: These help to keep blood flowing.

Heart transplant: A damaged or failing heart is replaced with a healthy heart from a deceased donor. This is reserved for people with severe heart failure.

Medical interventions: Taking statins, for instance, can reduce blood cholesterol levels and help to prevent arteries from becoming blocked.

Valve replacements: These can be mechanical (man-made) or biological (from cows or pigs). Faulty valves can be replaced to ensure blood flows through the heart in the right direction.

Cancer

Cancer occurs when cells in the body start to divide uncontrollably. This causes tumors. There are more than 200 different types of cancer. They can affect organs such as the lungs, tissues such as nervous tissue, and systems such as the immune system. Different cancers behave in different ways and require different treatments.

Cancer is the second leading cause of death globally. One in six deaths is due to cancer. In the US, one in two women and one in three men will develop cancer at some point in their lifetime.

TYPES OF CANCER
Tumors are either benign or malignant.

Benign tumors grow slowly and are not normally cancerous. They often grow within a membrane that limits their growth. They don't spread and are not usually dangerous, but if they become large and problematic, they may need to be removed.

Malignant tumors are cancerous cells that grow quickly and are not enclosed in a membrane. They can metastasize, or spread to other places in the body via the bloodstream and lymphatic system. The initial tumor is called the **primary tumor**. Metastatic tumors are called **secondary tumors**. Left untreated, malignant tumors can be very dangerous.

The primary tumor secretes chemicals into the surrounding environment.

The chemicals stimulate blood vessels to grow toward the tumor. They supply the tumor with oxygen and nutrients.

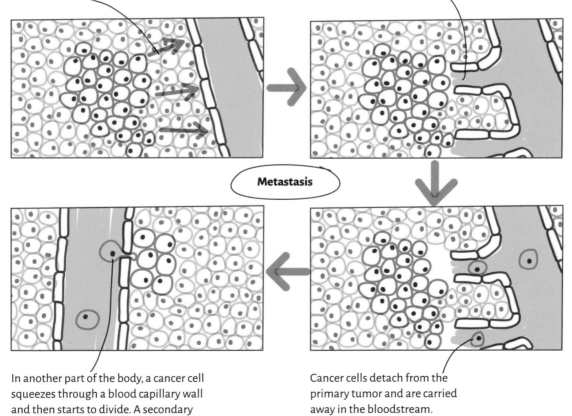

Metastasis

In another part of the body, a cancer cell squeezes through a blood capillary wall and then starts to divide. A secondary tumor begins to grow.

Cancer cells detach from the primary tumor and are carried away in the bloodstream.

CAUSES OF CANCER

Although the causes of many cancers are still poorly understood, scientists now realize that both genetic and environmental factors are important. The causes of cancer are complex and varied. Often multiple factors are involved.

Lifestyle factors: Smoking, obesity, and excessive alcohol consumption can increase the risk of developing cancer.

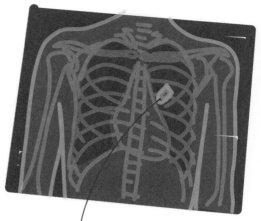

Carcinogens: These are chemicals that damage DNA, e.g. asbestos can cause some lung cancers.

Faulty genes: Some people with breast cancer carry a faulty version of a BRCA gene.

Ionizing radiation: Also damages DNA, e.g. too much UV light from the sun can cause skin cancer.

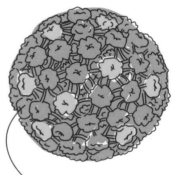

Viral infections: Around 15% of human cancers are caused by viral infections, e.g. human papillomavirus (HPV) can cause cervical cancer, so many teenagers are now vaccinated against this.

CANCER TREATMENTS

There are many different ways of treating cancer. No two cancers are the same, so doctors choose the most suitable treatment for each individual. The following are common treatments.

Surgery: Cancerous tissues can be removed by surgery.

Chemotherapy: Chemicals are used to stop cancer cells from dividing. The treatment is injected into the bloodstream. It has side effects because it also damages healthy dividing cells. This is why cancer patients sometimes lose their hair. Side effects like this are temporary because the healthy cells bounce back.

Radiotherapy: Uses focused beams of ionizing radiation, such as X-rays, to destroy cancer cells. Like chemotherapy, radiotherapy also affects normal cells. This can cause side effects in the treatment area, but they are usually temporary.

DRUGS AND DISEASES

Drugs are commonly used to treat diseases. Some drugs, such as painkillers, relieve the symptoms but don't cure the disease. Others tackle the underlying cause of the disease; for example, antibiotics kill the bacteria that cause infection.

Drug discovery

Many of the drugs that are used today were originally discovered in plants, animals, and microorganisms, but now scientists use a combination of chemistry and computer modeling to discover new medicines.

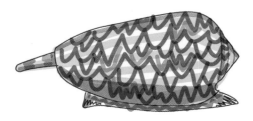

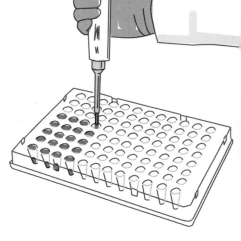

DRUGS FROM ANIMALS

A toxin from cone snails called ziconotide is used to treat people with severe, chronic pain. It is 1,000 times more powerful than morphine.

DRUGS FROM MICROORGANISMS

In 1928, Scottish scientist Alexander Fleming discovered a mold that could kill bacteria. The mold released a substance that he named penicillin. Ten years later, scientists extracted penicillin and showed it can cure bacterial infections in people. There are now many different types of penicillin, and they are used to treat many different infections.

DRUGS FROM COMPUTER MODELING

Scientists study the structure of useful drugs and then use computer modeling to determine how the structure could be tweaked to make the medicine better. Huge panels of synthetic molecules are then tested in the lab to see which work best. Promising molecules are developed further.

DRUGS FROM PLANTS

In medieval times, people chewed beaver tails to treat headaches. It worked because beavers eat willow trees, which contain an aspirin-like compound. In 1897, German chemist Felix Hoffmann synthesized acetyl salicylic acid, or "aspirin," for the first time. It is now one of the most widely used medications in the world.

Drug development

Drug development is costly and labor-intensive. There are three main stages:

Stage 1: During preclinical testing, drugs are tested on cultured cells and in computer models. Most never make it past this stage because they don't work or prove to be toxic.

Stage 2: The most promising candidates are tested on laboratory animals, such as rats and mice. The drugs' effectiveness and toxicity are assessed, and any side effects are noted.

Stage 3: Drugs that perform well on animal tests can then be tested on humans in clinical trials. There are three phases of clinical trials.

Number of compounds

500–1,000

10–20

5–10

2–5

1–2

Phase I: The drug is tested on a small numbers of healthy volunteers to check it is safe. Different doses are tried, and side effects are monitored.

Phase II: The drug is tested on larger numbers of ill people to see if it works. Most phase II trials are **randomized**: half of the patients receive the drug, and the other half receive a control. The **control** is either a **placebo**, which is an inactive version of the drug, or a standard treatment that is already known to work. Often these studies are **double blind**, which means that neither the researchers nor the patients know who has received the experimental drug until after the study is finished.

Phase III: If the drug looks promising, the phase II trial is repeated in larger groups of patients. If it is effective and well tolerated, the drug may be approved for widespread clinical use.

LIFESTYLE AND HEALTH

Diseases are complicated. Many factors contribute to disease, but most diseases are not inevitable. There are many things you can do to help stay healthy and disease-free.

Risk factors

A **risk factor** can be anything that increases the chances of developing a disease. Some risk factors are non-modifiable— they cannot be avoided. Others are modifiable and so can be changed.

Risk factors mainly apply to noncommunicable diseases, but they can also apply to communicable diseases.

For example, if someone is malnourished, their immune system may be weakened and they may be at heightened risk of viral infection.

Non-modifiable factors

Faulty genes, e.g. cystic fibrosis is caused by a fault in a single, inherited gene.

Gender, e.g. women are more likely to develop breast cancer than men.

Age, e.g. Alzheimer's disease mainly affects older people.

Modifiable factors: Lifestyle

Smoking is a risk factor for lung cancer and cardiovascular disease.

Obesity is a risk factor for diabetes, cardiovascular disease, and cancer.

Alcohol consumption is a risk factor for addiction, liver disease, and cardiovascular disease.

Physical inactivity is a risk factor for stroke, diabetes, and cardiovascular disease.

Unprotected sex is a risk factor for developing sexually transmitted diseases (STDs).

Modifiable factors: Environmental

Dirty water is a risk factor for diarrhea, cholera, and dysentery.

Exposure to pollutants, such as asbestos, is a risk factor for lung cancer and respiratory disorders.

Multiple risk factors can contribute to the emergence of a disease. For example, smoking, physical inactivity, poor diet, obesity, age, and a related family history, are all related risk factors for cardiovascular disease.

Just because there is a link or correlation between a risk factor and a disease, doesn't necessarily mean the risk factor causes the disease. Once a link is established, scientists do extra studies to establish the cause.

Staying healthy

There are many things we can do to help us stay healthy and reduce the risk of disease. These include regular exercise, not smoking, limiting alcohol intake, eating healthily, and maintaining a healthy weight.

Body Mass Index (BMI) is used to gauge if a person's weight falls in the normal range.

BMI	CATEGORY
Over 30	Obese
25–30	Overweight
19–25	Normal
Less than 19	Underweight

BMI = weight (kg) / square of height (m). Therefore, if a person weighs 40 kg (88 lb.) and is 1.5 m (4ft 9 in) tall, the BMI is 40 / 2.25 = 17.7. This is slightly underweight.

Nutrition

It's important to eat a healthy diet to ensure that the body gets all the nutrients it needs to operate efficiently.

Fiber: Found in vegetables, bran, and nuts; helps to keep food moving through the gut.

Carbohydrates: Found in cereals, potatoes, pasta, bread, and rice; a good source of energy.

Lipids (fats and oils):Found in butter, oil, and nuts; provide insulation; a useful source of energy that can be stored.

Vitamins: Found in fruit, vegetables, and dairy products; required in small amounts to help cells function properly.

Proteins: Found in meat, fish, eggs, beans, pulses, and dairy products; used for growth and repair.

Minerals: Found in salt, milk (calcium), and liver (iron); required in small amounts to help cells function properly.

Water: Found in water, fruit juice, and milk; needed to maintain bodily fluids and keep cells healthy.

RECAP

BODY MASS INDEX

Used to gauge if weight falls in normal range. BMI = weight / height squared.

HEALTH INEQUALITIES

Avoidable health differences between groups of people. Influenced by geography, income, age, gender, and education.

NUTRITION

Eat a balanced diet including carbohydrates, proteins, vitamins, and minerals.

HUMAN HEALTH AND DISEASE

RISK FACTORS

Include obesity and smoking. Interact with one another. Make disease more likely. Don't necessarily cause disease.

Preclinical and animal testing lead to extensive clinical trials. Human tests are randomized and double blind.

DEVELOPMENT

DRUGS

DISCOVERY

Historically from plants, microorganisms, and animals. Increasingly by computer modeling.

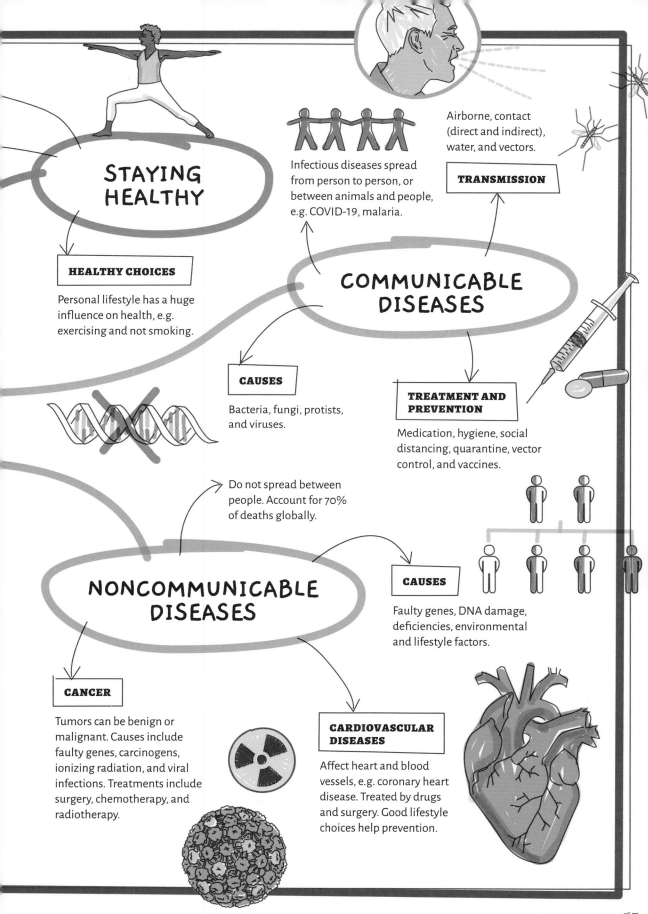

STAYING HEALTHY

HEALTHY CHOICES

Personal lifestyle has a huge influence on health, e.g. exercising and not smoking.

Infectious diseases spread from person to person, or between animals and people, e.g. COVID-19, malaria.

TRANSMISSION

Airborne, contact (direct and indirect), water, and vectors.

COMMUNICABLE DISEASES

CAUSES

Bacteria, fungi, protists, and viruses.

TREATMENT AND PREVENTION

Medication, hygiene, social distancing, quarantine, vector control, and vaccines.

Do not spread between people. Account for 70% of deaths globally.

NONCOMMUNICABLE DISEASES

CAUSES

Faulty genes, DNA damage, deficiencies, environmental and lifestyle factors.

CANCER

Tumors can be benign or malignant. Causes include faulty genes, carcinogens, ionizing radiation, and viral infections. Treatments include surgery, chemotherapy, and radiotherapy.

CARDIOVASCULAR DISEASES

Affect heart and blood vessels, e.g. coronary heart disease. Treated by drugs and surgery. Good lifestyle choices help prevention.

CHAPTER 10

ECOLOGY

Living things do not live in isolation. They interact with one another and with their physical environment. Ecology is the study of these ecosystems. It teaches us how all life is interconnected and how changes to one part of an ecosystem can have profound repercussions elsewhere. Living things exist in complex food webs where competition is rife, but specialized features, known as adaptations, enable living organisms to survive and reproduce. Meanwhile, important chemicals, such as nitrogen and carbon, are recycled globally, ensuring that life can persist. Let's delve into some of these relationships and recycling schemes.

ECOSYSTEMS

All living things are part of an ecosystem. An ecosystem is made up of communities of living things and the physical environment in which they exist. There are many different types of ecosystems, for example, coral reefs, deserts, cities, and prairies.

A **community** is made up of two or more populations of organisms. A **population** is all the members of a species that live in a single geographical area. An ecosystem is the interaction between two or more different populations and their environment.

Amazon rain forest

Ecosystems are like Russian dolls because they have smaller and smaller ecosystems nestled inside them.

Forests contain trees. Each tree is its own ecosystem. It depends on its physical environment for sunlight, water, and nutrients, and it interacts with the organisms that live in and on it, such as birds and insects.

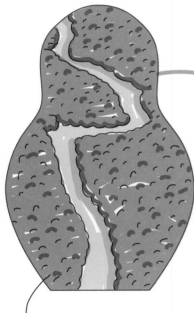

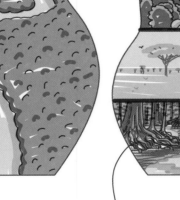

The Amazon rain forest occupies c. 5.5 million km² (c. 2.1 million miles²). It is the world's largest tropical rain forest. It can be considered as an ecosystem.

The Amazon rain forest is far more than just forest. It contains many different habitats, such as savanna and swamps. These are also considered as ecosystems.

Leaf-cutter ants chew off bits of leaf and then carry them back to their nest. Each ant is its own ecosystem. It depends on its physical environment for warmth and food, and it interacts with the organisms that interact with it, e.g. the birds and reptiles that eat the ants, and the bacteria that live inside the ant's gut.

iNTERDEPENDENCE

All life is connected, and all of the organisms that live in an ecosystem depend on one another. This is called **interdependence**. It means that even small changes to one part of an ecosystem can have big repercussions elsewhere.

Plants depend on animals. Animals act as pollinators and seed dispersers. They provide plants with nutrients via their feces, and when animals die and decompose, they fertilize the ground and help the plants to grow.

Animals depend on plants. Plants provide a source of food, shelter, and raw materials for building. They also produce the oxygen that animals need to breathe.

Food chains

Food chains describe how living things acquire their food, and how nutrients and energy pass from one organism to the next.

Sea otters, for example, share a simple food chain with sea urchins and kelp. Food chains are made of producers and consumers.

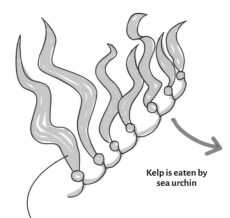

Kelp is eaten by sea urchin

Sea urchin is eaten by otter

Kelp is a **producer**. Producers are the plants at the bottom of a food chain. Using energy from the sun, they use photosynthesis to convert water and carbon dioxide into glucose. This fuels the plant's growth and provides a source of nutrients for the organisms higher up the chain.

Sea urchins are **consumers**. Consumers are the organisms that eat other things. Here, sea urchins are the **primary consumer.**

Sea otters are another consumer. Because they feed on sea urchins, they are the **secondary consumer**.

Animals that hunt and kill other animals are called **predators**. Those that are hunted and killed are called **prey**.

Food webs

In reality, life is never as simple as a single food chain. Instead, food chains link together to form more complicated **food webs**. Food webs illustrate how the various members are all interconnected.

Changes to any part of the food web leads to changes in other areas. Sea otters, sea urchins, and kelp form a food chain, but this is just a small fragment of a much bigger food web.

In this diagram, sea otters are **apex predators** because they are at the top of the food web.

Some predators eat lots of different things. In this ecosystem, crabs eat little fish, bivalves, and plant matter. This is a useful strategy because it means if one food source declines, they can eat something else. This can also result in causal effects higher up the food web, e.g. if crabs eat more bivalves, then the sea stars may struggle, and their numbers may decline.

Some organisms have restricted diets and only eat one thing. Bivalves are a type of mollusk. They are filter feeders so they eat just plankton. This is a risky strategy because if plankton is in short supply, the bivalves may starve. If the bivalves decline, the crabs and small predatory fish that feed on them may also struggle, and this in turn can have progressive effects higher up the food web.

ENVIRONMENTAL INFLUENCES ON LIFE

Living things survive in a great variety of habitats, where they are affected by the conditions they experience. Changes in **biotic** (living) and **abiotic** (nonliving) factors have a big influence on individual organisms and the communities they live in.

Biotic factors

Food: Living things need food to provide them with energy. When there is plenty of food, organisms can grow and reproduce successfully. When food is scarce, survival and reproduction are compromised. The fish in this pond are doing well because there are lots of bugs and larvae to eat.

Pathogens: Living things are susceptible to pathogens and the diseases they carry. If a population has never encountered a particular pathogen before, they will have no immunity to it and the effects can be deadly. If your garden has a pond, it's never a good idea to introduce tadpoles from elsewhere. They may carry disease.

Predators: If a new predator arrives in an environment, it can radically change the ecosystem. If a new domestic cat moves into the neighborhood, it could decimate the fish in this pond.

Competition: Sometimes species compete with one another for resources, such as food and territory, e.g. the lilies in this pond are competing for nutrients and space.

Abiotic factors

Water: Animals and plants need water to survive. Some organisms, such as cacti and camels, have adapted to survive with low water levels, while others, such as fish and water lilies, have adapted to life in a watery environment.

Light: Animals and plants both respond to light. Plants need light for photosynthesis, so they don't grow well in dark environments such as caves or the sea floor. Some plants are adapted to grow in bright light, while others thrive in the shade. Light intensity also affects animals, e.g. frogs lay their eggs when the days lengthen in spring.

Soil pH and mineral content: The types of plant that grow are influenced by soil pH. The soil in this garden is alkaline, so lavenders grow well. Minerals are also important. Plants need nitrates to help make proteins, and magnesium to help make chlorophyll.

Temperature: Plants don't grow well at low temperatures, e,g. in the cold Arctic, plants are stunted and small. This presents a problem for the animals that feed on them, so their numbers are limited in cold areas. In turn, this affects the number of carnivores that can survive. In this pond, growth is limited during the winter, when the temperature is low.

Gas levels: Plants need carbon dioxide for photosynthesis, and animals need oxygen for aerobic respiration, e.g. fish are very sensitive to oxygen levels in the water.

Wind: Plants are affected by wind intensity and direction. Plants that have evolved to live in windy places may have acquired useful features. The ornamental grasses in this garden have small, narrow leaves that help reduce moisture loss during transpiration.

ADAPTATIONS

Living things have developed features that help them to survive. These are called **adaptations**. Adaptations enable organisms to live in Earth's many, diverse ecosystems.

Types of adaptation

There are three main types of adaptation:

★ **Structural**: Physical features, such as the size, shape, or coloring of an organism.

★ **Behavioral**: Responses made by an organism that help it survive, e.g. mating, migrating, or searching for food.

★ **Functional**: Bodily processes that help an organism to survive, e.g. changes in metabolism or the ability to produce novel molecules.

SPOTLIGHT ON ANIMAL ADAPTATIONS: POLAR BEAR
Polar bears live in the Arctic where it is bitterly cold. They have many special adaptations.

Polar bears can hold their breath for up to three minutes under water. This helps them to sneak up on their prey underwater (functional).

Polar bears dig dens in the snow where they give birth and then nurse their cubs (behavioral).

Mother polar bears can go without food for more than eight months. This enables them to stay in their den during pregnancy and when they are nursing their young (functional).

Thick layers of body fat provide insulation (structural).

A low surface area to volume ratio helps to retain heat (structural).

Long, curved, sharp claws help them to kill and eat their prey (structural).

Small bumps on their footpads, called papillae, give traction on the ice (structural).

Polar bears are strong swimmers. This helps them to pursue and catch their prey (behavioral).

SPOTLIGHT ON PLANT ADAPTATIONS: BROMELIAD

Most plants acquire their nutrients and water from the soil via their roots, but some bromeliad plants do it differently. They grow on other plants, and their roots are not connected to the soil.

The water attracts tiny insects that become food for the tadpoles. Tadpole droppings and decaying insects provide nutrients for the plant.

Brightly colored flowers protruding from a tall stem attract pollinators, such as moths and hummingbirds (structural).

The base of the leaves forms a tiny pond that collects rainwater. Tree frogs lay their eggs in the pond, and the tadpoles develop here (structural).

The leaves are covered in fine hairs that rapidly absorb any rainwater that falls on them (structural).

Many bromeliads grow high up in tropical tree canopies where they receive bright sunlight for photosynthesis.

Some bromeliads use an unusual form of photosynthesis. It lets them close their stomata during the day to reduce water loss, and then open them at night to obtain carbon dioxide. This is the opposite of most plants

Extremophile adaptations

Organisms that survive in extreme environments are called **extremophiles**. They often have unusual adaptations.

★ An Antarctic fish called the shorthorn sculpin produces antifreeze proteins that enable it to live in the icy cold waters.

★ Some bacteria live in extremely salty environments, such as the Dead Sea. Their cells are adapted to retain water, and not lose it via osmosis.

★ The kangaroo rat survives in hot deserts without drinking by avoiding the sun, producing superconcentrated urine, and reabsorbing as much water as possible.

COMPETITION

Species interact with one another in many different ways. Often, they compete with one another for resources. When members of different species compete, it is called **interspecific competition**. When members of the same species compete, it is called **intraspecific competition**. Sometimes the interactions are mutually beneficial, and sometimes they are one-sided.

Competition in plants

Plants compete for light, water, space, and nutrients. Big, tall plants, for example, use a lot of resources, such as water and minerals. This makes it difficult for smaller plants to grow in their shade. Unlike animals, plants can't fight one another or run away, so they have evolved a range of adaptations to help them compete.

Growth: Some plants have big leaves with large surface areas, which help to maximize photosynthesis. Others, such as this buddleia, have smaller leaves but grow long and straggly. This helps the plant to access light.

Flowering: Some plants avoid competition with one another by flowering at different times, e.g. daisies and many other flowers bloom in late spring and summer when the days are longer and there is more sunlight available for photosynthesis. Some plants, such as the tobacco plant (*Nicotiana*), have even evolved to flower at night, when they are pollinated by nocturnal insects, such as moths.

Seed dispersal: Some plants have specialized seeds that can be transported far and wide by the wind or by other animals, e.g. the light, fluffy seeds of the dandelion are transported by the wind.

Chemical warfare: Some plants defend their territory by releasing toxic chemicals into the soil. Others, such as the stinging nettle, have evolved formic acid-containing stinging cells that help deter herbivores from eating them.

Competition in animals

Animals compete for territory, food, water, and mates. Sometimes competition is relatively peaceful, such as when birds sing to defend their territory. Sometimes it is confrontational and violent; for example, meerkats from one group will sometimes kill individuals from a different group in order to protect their patch.

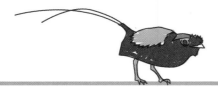

COMPETITION FOR MATES

Some animals fight to gain access to mates, e.g. male red deer will lock antlers and fight. The victor then gets access to the females.

Other animals have a more peaceful solution, e.g. male manakins perform intricate dances and displays to impress the females. The best dancer gets to mate.

COMPETITION FOR FOOD

This is very common, e.g. lions and spotted hyenas compete with each other on the plains of Africa. They attack each other, steal from each other, and sometimes kill each other's young.

Some animals only eat one food type, e.g. pandas eat mainly bamboo. Being a fussy eater is a risky strategy because if the food becomes scarce, life can be difficult.

Other animals eat a variety of different foods, e.g. coyotes eat a broad diet that includes mice, voles, birds, and plants. Animals with broader diets are more likely to survive during lean periods.

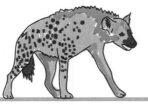

COMPETITION FOR TERRITORY

Territories contain important resources, such as mates, food, and water, so some animals defend them vigorously, e.g. a male elephant seal will defend the patch of beach that contains his personal harem of females. If he does this successfully, he will get to mate with all of the females.

Parasitism and mutualism

Parasitism is another form of interaction. **Parasites** feed on or in another living organism, which is called the **host**. The parasite takes what it needs from the host, but the host gets no benefit from this interaction. For example, lice, mosquitoes, fleas, and vampire bats are all parasites.

The tongue-eating louse enters a fish through its gills. It attaches to the tongue where it severs the blood vessels, causing the tongue to fall off. It then replaces the tongue and lives in the fish's mouth where it feeds on the host's blood.

Tapeworms live in the digestive system of vertebrates. They have no digestive system of their own, so they absorb all the nutrients they need from their host. Tapeworms in human can grow up to 15 meters (49 feet) in length. Eggs released in the host's feces can then infect other hosts.

Sometimes both species benefit from an intense and close relationship. This is called **mutualism**.

When bees visit flowers, they obtain food in the form of nectar. Pollen sticks to the bees, which they then transfer to other plants. The bees help the flowering plants to reproduce.

The eyelight fish has special organs under its eyes that are filled with bioluminescent bacteria. The bacteria receive nutrients from the fish, and the fish uses the light to attract prey and signal to mates.

KEYSTONE SPECIES

Some organisms exert a particularly large influence on the environment. They are called **keystone species**. They are important because they create opportunities for other species to thrive. **Biodiversity** is the variety of different species that live in a particular habitat. Keystone species increase biodiversity. When they disappear from an ecosystem, the ecosystem either disappears or changes radically. Wolves, elephants, and sea otters are all keystone species.

The wolves of Yellowstone National Park

Conservationists realize the value of keystone species. The wolves of Yellowstone National Park show how sometimes, introducing a single keystone species is all it takes to boost diversity.

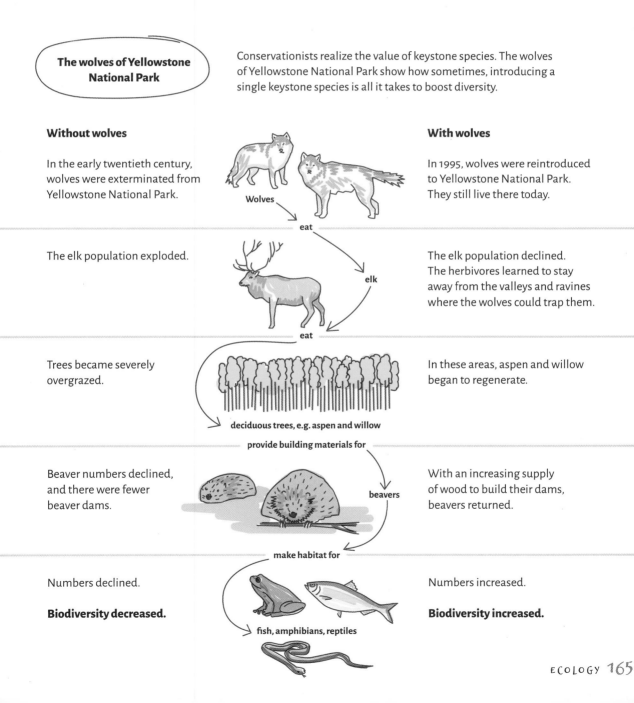

Without wolves

In the early twentieth century, wolves were exterminated from Yellowstone National Park.

Wolves

eat

With wolves

In 1995, wolves were reintroduced to Yellowstone National Park. They still live there today.

The elk population exploded.

elk

The elk population declined. The herbivores learned to stay away from the valleys and ravines where the wolves could trap them.

eat

Trees became severely overgrazed.

deciduous trees, e.g. aspen and willow

In these areas, aspen and willow began to regenerate.

provide building materials for

Beaver numbers declined, and there were fewer beaver dams.

beavers

With an increasing supply of wood to build their dams, beavers returned.

make habitat for

Numbers declined.

Biodiversity decreased.

fish, amphibians, reptiles

Numbers increased.

Biodiversity increased.

GLOBAL RECYCLING OF ELEMENTS

All living things are made from the same basic building blocks, such as carbon, nitrogen, oxygen, and hydrogen. There is a finite amount of these resources, so they are constantly recycled between living things and the environment. Carbon, water, and nitrogen are recycled in separate schemes.

Carbon cycle

The carbon cycle is the global process that recycles carbon atoms. Carbon flows from the atmosphere into living things and the earth, and then back into the atmosphere.

Carbon dioxide in the air

Photosynthesis: Sunlight powers the conversion of carbon dioxide and water to glucose and oxygen. Carbon is removed from the environment.

Animal respiration: Carbon is returned to the environment as carbon dioxide.

Carbon compounds in green plants

Plant respiration: Glucose and oxygen are converted into carbon dioxide, water, and energy. Carbon is returned to the environment.

Carbon compounds in animals

Feeding: Carbon locked in plants is transferred to animals when the plants are eaten.

Death and decay

Decay

Death and decay are an important component of global recycling schemes. When living things die, they decompose and return useful chemicals, such as carbon and nitrogen, to the environment. Living organisms, such as maggots, fungi, and bacteria, oversee this process. They are known as **decomposers**.

Combustion: Fossil fuels are burned, releasing carbon dioxide and energy. Carbon is returned to the environment as carbon dioxide.

Carbon compounds in fossil fuels

Fossil fuel production: Over millions of years, dead plants and animals are converted to coal and oil. These fossil fuels are rich in carbon.

Microorganism respiration: Carbon is returned to the environment as carbon dioxide.

Death and decay

Carbon compounds in dead organisms

THE RATE OF DECAY

Living things decompose and decay at different rates. For example, a dead body could last for centuries inside a cool, dark cave, but the carcass of a baby lion will be reduced to bones within weeks if it is left exposed on the hot African plain. The rate of decay is influenced by various factors including: temperature, water availability, and oxygen availability.

Temperature

Decomposers use enzymes to break bigger molecules into smaller ones. Warmer temperatures speed up the rate of decomposition because they increase enzyme activity. However, if the temperature is too high, the enzymes lose their precise 3D shape and are unable to bind to their substrate. Decay slows down or stops. Really cold temperatures also slow down decay.

Water availability

Decay occurs more quickly in moist environments. This is because decomposers need water to help them digest their food, and to prevent them from drying out.

Oxygen availability

Most decomposers respire aerobically. This means they need oxygen to help break down their food and fuel the decomposition process. As a result, decomposition occurs more quickly in oxygen-rich environments.

Water cycle

The water cycle is the global process that recycles water. It provides fresh water for land-living animals and plants, before draining into the rivers and oceans of the world and then being returned to the atmosphere.

Precipitation: The wind blows the clouds over the land, and the water droplets return to earth in the form of rain, snow, hail, or sleet.

Condensation: As the moist air rises, it cools and condenses into liquid water droplets that form clouds.

Transpiration and respiration: Plants draw up water from the ground via transpiration. As water evaporates from plant leaves, more water is drawn into the roots and up into the plant. When plants respire, they return water vapor to the atmosphere.

Runoff: Some water runs into streams and rivers and is returned to the sea.

Water vapor

Evaporation: The sun warms the Earth's surface. Water is turned from a liquid into a vapor which then rises into the air.

Ocean

Respiration: When animals respire, they also return water vapor to the atmosphere. Animals also return water to the environment in their urine, feces, and sweat (mammals only).

Shallow storage: Some water remains in shallow layers of soil where it is available to plants.

Deep storage: Water seeps into the ground through gaps in soils and rocks. Some is stored in underground rocks called **aquifers**.

Nitrogen cycle

The nitrogen cycle is the global process that recycles nitrogen. Nitrogen is important because it is needed by living things to help make proteins and DNA. Although nitrogen gas makes up 80% of Earth's atmosphere, living things cannot use it in this form. The nitrogen cycle converts atmospheric nitrogen into other forms that can be used by living things, and recycles nitrogen molecules through organisms and the environment.

Denitrification: Nitrates are converted into nitrogen gas. This is done by denitrifying bacteria. Denitrifying bacteria are anaerobic so they don't need oxygen. This means they do well in waterlogged soil.

Nitrogen in the atmosphere

Nitrogen uptake: Plants absorb nitrates from the soil. They use them to make proteins and help them grow. This nitrogen passes up the food chain when plants are eaten by animals.

Nitrogen-fixing bacteria in root nodules of legumes

Denitrifying bacteria

Nitrogen uptake

Decomposers (fungi and bacteria)

Decomposition: When organisms die and decompose, the proteins in their bodies are broken down by bacteria and fungi to form ammonia. Animal excrement is also broken down to form ammonia.

Nitrates

Nitrogen-fixing bacteria in the soil

Nitrifying bacteria

Nitrites

Ammonia

Nitrogen fixation: Nitrogen gas is converted into a form that can be used by living things. This is done by nitrogen-fixing bacteria. Nitrogen-fixing bacteria are aerobic so they need oxygen.

Nitrogen-fixing bacteria can be found in the soil and in the roots of leguminous plants, such as peas, beans, and clover.

Nitrification: Ammonia is converted into nitrites and then to nitrates. This is done by nitrifying bacteria. Nitrifying bacteria are aerobic so they need oxygen.

RECAP

Living things have an effect, e.g. competition, food, pathogens, and predators.

BIOTIC

Nonliving things have an effect, e.g. water, light, gas levels, pH, minerals, temperature, and wind.

ABIOTIC

Communities of living things and their physical surroundings.

ENVIRONMENTAL INFLUENCES ON LIFE

ECOLOGY

Carbon is recycled between the atmosphere, the earth, and living things.

CARBON CYCLE

Atmospheric nitrogen is converted into usable forms, and back again.

NITROGEN CYCLE

Responses made by organisms, e.g. migrating and foraging.

BEHAVIORAL

GLOBAL RECYCLING

WATER CYCLE

Water journeys between the sky, land, and oceans.

STRUCTURAL

Physical features, e.g. size, shape, and coloring.

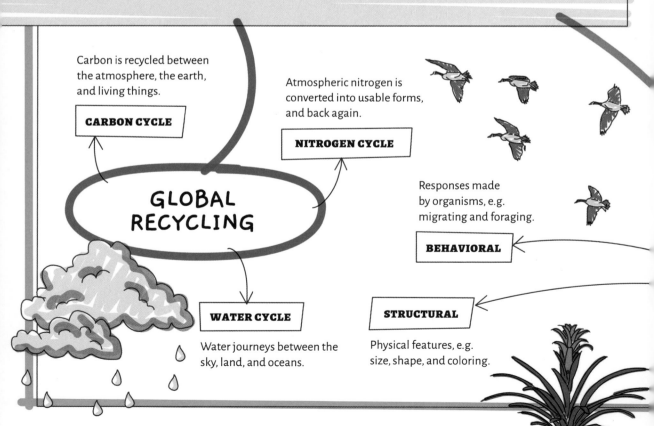

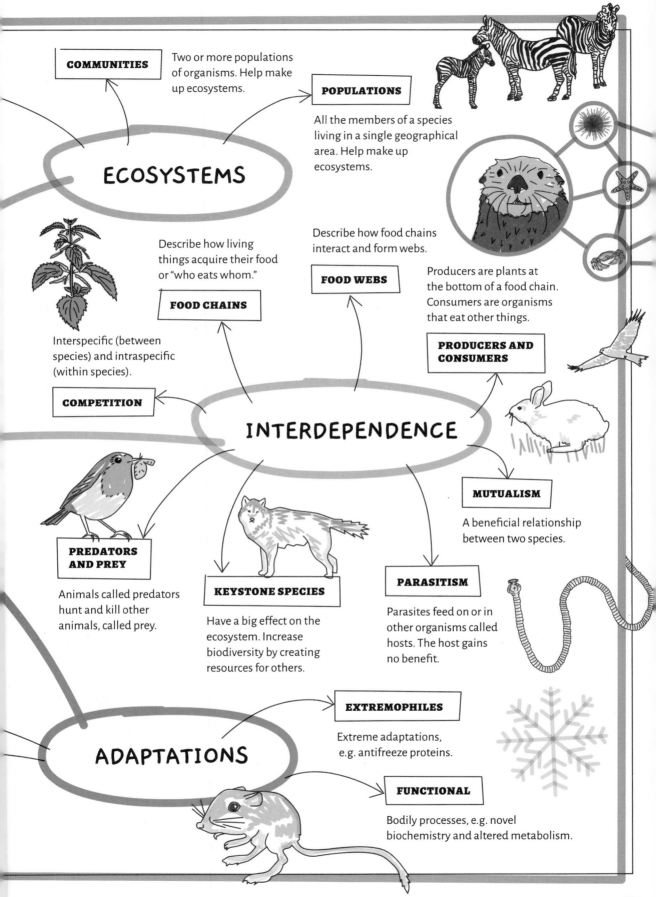

COMMUNITIES Two or more populations of organisms. Help make up ecosystems.

POPULATIONS

All the members of a species living in a single geographical area. Help make up ecosystems.

ECOSYSTEMS

Describe how living things acquire their food or "who eats whom."

Describe how food chains interact and form webs.

FOOD WEBS

FOOD CHAINS

Producers are plants at the bottom of a food chain. Consumers are organisms that eat other things.

Interspecific (between species) and intraspecific (within species).

PRODUCERS AND CONSUMERS

COMPETITION

INTERDEPENDENCE

MUTUALISM

A beneficial relationship between two species.

PREDATORS AND PREY

Animals called predators hunt and kill other animals, called prey.

KEYSTONE SPECIES

Have a big effect on the ecosystem. Increase biodiversity by creating resources for others.

PARASITISM

Parasites feed on or in other organisms called hosts. The host gains no benefit.

EXTREMOPHILES

Extreme adaptations, e.g. antifreeze proteins.

ADAPTATIONS

FUNCTIONAL

Bodily processes, e.g. novel biochemistry and altered metabolism.

CHAPTER 11

BIOLOGY IN THE TWENTY-FIRST CENTURY

Earth is now very different from how it was even a few hundred years ago. Forests have been razed, the global recycling of elements has become disrupted, massive amounts of carbon dioxide have been pumped into the atmosphere, and vast swathes of land have been repurposed for agriculture. This comes at a cost. Now the planet is warming, species are being driven to extinction, and the ecosystems that humans rely on for survival are increasingly at risk. In this chapter, you'll explore these issues, and learn about some of the solutions that can help to save the planet.

THE ANTHROPOCENE

Our planet is home to more than 7 billion people, and the number continues to grow. Human activity has changed Earth beyond all recognition. Now scientists think it has become so different that the current geological time period deserves a new name. They want to call it the **Anthropocene**, which means "the age of humans."

Ten thousand years ago, there were no farms or cities. Now two-thirds of the world's ice-free land is devoted to human uses.

The way we manage our land is having profound repercussions. Half of Earth's mature tropical rain forests—around 8 million km² (3 million miles²)—have been destroyed.

Rain forests produce oxygen, and help to maintain the world's water cycle by adding water to the atmosphere via transpiration.

They are also **carbon sinks**. This means they absorb carbon dioxide from the atmosphere. As humans continue to pump large volumes of carbon dioxide into the atmosphere, we lose this resource at our peril.

Deforestation and habitat destruction are both major drivers of biodiversity loss. Plant species are disappearing, and animals are losing their habitats. This is pushing the surviving animals into closer proximity with humans, which is leading to conflict.

Scientists also think it increases the likelihood of animal diseases jumping species and infecting humans.

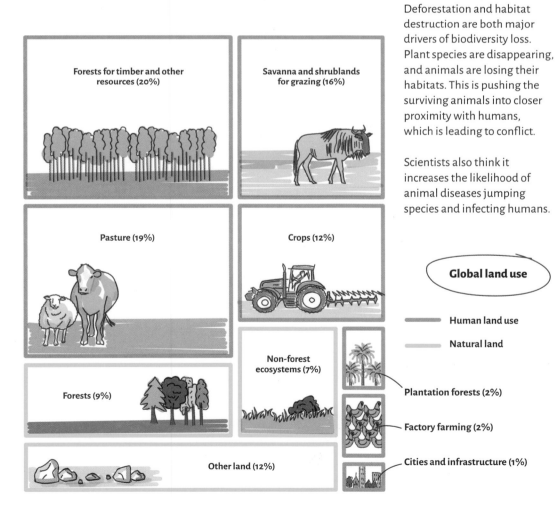

Forests for timber and other resources (20%)

Savanna and shrublands for grazing (16%)

Pasture (19%)

Crops (12%)

Forests (9%)

Non-forest ecosystems (7%)

Other land (12%)

Global land use

— Human land use

— Natural land

Plantation forests (2%)

Factory farming (2%)

Cities and infrastructure (1%)

PLANETARY BOUNDARIES

Scientists have highlighted nine important environmental processes, including climate change and biodiversity loss. To keep the planet healthy and habitable, we need to keep these processes within a particular boundary or "safe zone," but now four of these boundaries have been breached. It will take a concerted global effort to help keep our planet habitable.

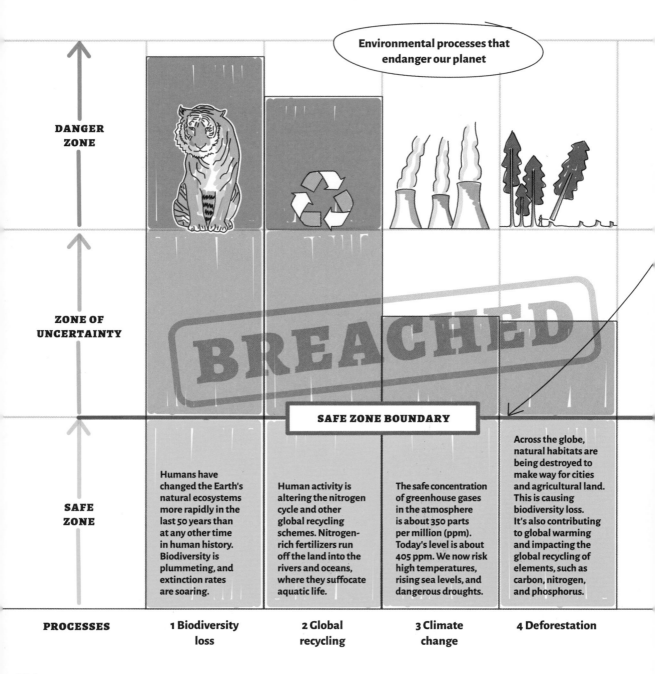

Environmental processes that endanger our planet

DANGER ZONE

ZONE OF UNCERTAINTY

BREACHED

SAFE ZONE BOUNDARY

SAFE ZONE

Humans have changed the Earth's natural ecosystems more rapidly in the last 50 years than at any other time in human history. Biodiversity is plummeting, and extinction rates are soaring.

Human activity is altering the nitrogen cycle and other global recycling schemes. Nitrogen-rich fertilizers run off the land into the rivers and oceans, where they suffocate aquatic life.

The safe concentration of greenhouse gases in the atmosphere is about 350 parts per million (ppm). Today's level is about 405 ppm. We now risk high temperatures, rising sea levels, and dangerous droughts.

Across the globe, natural habitats are being destroyed to make way for cities and agricultural land. This is causing biodiversity loss. It's also contributing to global warming and impacting the global recycling of elements, such as carbon, nitrogen, and phosphorus.

| PROCESSES | 1 Biodiversity loss | 2 Global recycling | 3 Climate change | 4 Deforestation |

It's important to understand these boundaries. Knowing about them can help us to rein in dangerous activities, prepare for future change, and prioritize resources.

All of these processes are interconnected, so a change to one process can have repercussions elsewhere. For example, climate change is leading to ocean acidification, which is endangering marine life and contributing to biodiversity loss.

Beyond this point we risk putting the planet in danger.

The boundaries for these processes are currently unquantified.

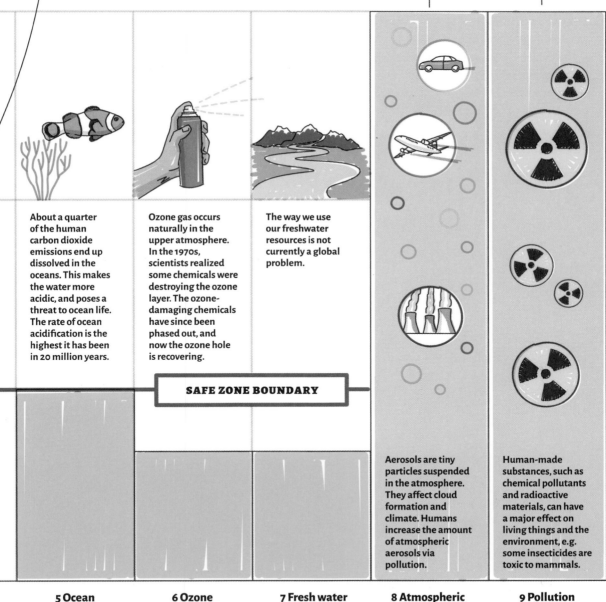

About a quarter of the human carbon dioxide emissions end up dissolved in the oceans. This makes the water more acidic, and poses a threat to ocean life. The rate of ocean acidification is the highest it has been in 20 million years.

Ozone gas occurs naturally in the upper atmosphere. In the 1970s, scientists realized some chemicals were destroying the ozone layer. The ozone-damaging chemicals have since been phased out, and now the ozone hole is recovering.

The way we use our freshwater resources is not currently a global problem.

SAFE ZONE BOUNDARY

Aerosols are tiny particles suspended in the atmosphere. They affect cloud formation and climate. Humans increase the amount of atmospheric aerosols via pollution.

Human-made substances, such as chemical pollutants and radioactive materials, can have a major effect on living things and the environment, e.g. some insecticides are toxic to mammals.

5 Ocean acidification

6 Ozone depletion

7 Fresh water use

8 Atmospheric aerosols

9 Pollution

THE GREENHOUSE EFFECT

The **greenhouse effect** is a process that occurs when gases in Earth's atmosphere trap heat from the sun, just like the glass in a greenhouse also traps heat. For millions of years, this was a good thing because the greenhouse effect helped to warm Earth and create the conditions for life to thrive. Now, however, human activities are changing this natural greenhouse effect. As we pump carbon dioxide and other greenhouse gases into the atmosphere, Earth's atmosphere is trapping more and more heat. This is leading to global warming.

1
Sunlight

Reflected sunlight

3

4
Reflected infrared radiation

5
Trapped infrared radiation

Infrared radiation

2

SPACE

Space is everything in the universe beyond the top of Earth's atmosphere. There are no gases here, so space is a vacuum. The distance from our sun to the top of Earth's atmosphere is around 150 million km (93 million miles).

ATMOSPHERE

The **atmosphere** is the layer of gases that surrounds the earth. It is about 480 km (300 miles) thick.

Preindustrial world

In preindustrial times, the concentration of greenhouse gases in the atmosphere was relatively stable. Although the global temperature varied a bit, overall the planet's temperature was quite steady.

EARTH

1 The sun is Earth's major source of energy. Sunlight travels through space and passes through the Earth's atmosphere.

2 When the sunlight reaches Earth, it warms our planet's surface. The heat that is created is in the form of infrared radiation.

3 Clouds in Earth's atmosphere reflect some of the sun's rays back into space.

4 Earth's surface reflects infrared radiation back into space.

5 Greenhouse gases trap some of the infrared radiation that is reflected from the earth. This makes the world warmer than it would be otherwise.

6 As the concentration of greenhouse gases increases in our atmosphere, less infrared radiation is being reflected into space, and more is becoming trapped in the atmosphere. This is raising the temperature of Earth's surface.

Reflected sunlight

3

6

Less reflected infrared radiation

6

More trapped infrared radiation

Industrial world

After the Industrial Revolution, the demand for fossil fuels skyrocketed. When they are burned for power, fossil fuels release carbon dioxide into the atmosphere. This has greatly increased the concentration of atmospheric greenhouse gases, leading to global warming and the current climate crisis.

CLIMATE CHANGE

Over the last 200 years, atmospheric carbon dioxide concentrations have increased by a third. This is disrupting the carbon cycle, exacerbating the greenhouse effect, and leading to climate change.

Sources of greenhouse gases

The main greenhouse gases in Earth's atmosphere are carbon dioxide, methane, nitrous oxide, and ozone. These are generated by natural processes, such as volcanic activity, and by many different human processes.

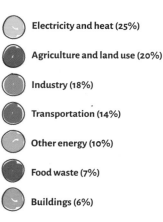

- Electricity and heat (25%)
- Agriculture and land use (20%)
- Industry (18%)
- Transportation (14%)
- Other energy (10%)
- Food waste (7%)
- Buildings (6%)

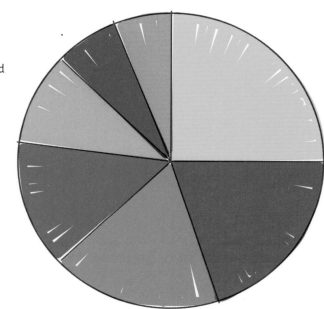

Evidence for climate change

Thermometer readings: The 20 warmest years on record have all been in the past 22 years. The hottest four years were 2015 to 2018.

Melting ice and sea level rise: Glaciers have been melting rapidly since the early 1900s. This is causing sea levels to rise. During the last 100 years, the average global sea level has risen by 19 cm (7.5 inches).

Ice cores: They provide an annual temperature record going back hundreds of thousands of years. Chemical analyses of ice cores confirm rapid recent warming.

Changes in seasonality: In some parts of the world, for example, spring is arriving earlier and winters are becoming milder.

Ocean acidification: The oceans are becoming more acidic as they absorb more carbon dioxide. Acidification of surface waters has increased by about 30% since the start of the Industrial Revolution.

Consequences of climate change

The world is now around 1°C (1.8°F) warmer than it was during preindustrial times, and the global temperature continues to rise. Small temperature changes like this can have enormous environmental effects.

Conflicts: As changing weather patterns and limited resources make life difficult, conflict could become more common. The 2007 conflict in Sudan's Darfur region has been described as the first climate change conflict.

Food and water: Droughts and poor crop yields will cause water and food shortages in many areas.

Rising sea levels: Sea levels could rise by more than 2 meters (6.5 feet) by 2100.

Melting ice: Arctic ice is 65% thinner than it was 45 years ago.

Agriculture: Although some regions will find it easier to grow crops, many places will experience desertification, resulting in declining crop yields.

Extreme weather: More heatwaves, droughts, and floods. More intense tropical storms.

Effects on the environment

Effects on people

Wildfires: Increasing in frequency.

Disease: As people and pathogens move around the planet, disease patterns will change, e.g. an extra 280 million people could be at risk from malaria.

Ecosystems: Are experiencing radical changes, e.g. if the world warms by more than 3°C (5.4°F), most of the world's coral reefs will disappear.

Migration: Climate change could lead to the displacement and migration of more than 140 million people in the next 30 years.

Biodiversity loss: 5% of all species could go extinct if warming exceeds 2°C (3.6°F) above preindustrial levels.

Flooding: 300 million people will experience flooding at least once a year unless carbon emissions are significantly reduced.

Dealing with climate change

Although it can seem overwhelming, there are many things that can be done to limit the amount of climate change. These include activities at the global, political, social, and personal level.

Geoengineering projects: These aim to modify Earth's climate system. One option is to suck carbon dioxide from the atmosphere by making artificial trees. Another is to use large mirrors to reflect more of the sun's energy back into space. These methods are controversial and still very much in development.

Alternative energy: Clean energy sources, such as solar, wind, and tidal, are increasingly replacing fossil fuels. Many cars now run on electricity, and many countries are phasing out the sale of gasoline and diesel cars.

Carbon capture: This is a developing technology. It involves capturing carbon dioxide from the waste gases emitted from power plants, and then storing it safely away underground. This prevents carbon dioxide from building up in the atmosphere.

Food production: Most of the world's 70 billion farm animals are intensively raised in factory farms. Factory farming releases vast quantities of greenhouse gases into the atmosphere, so now many people are advocating a move toward smaller-scale, more sustainable farming practices.

Adaptation strategies: These seek to limit the negative effects of climate change, rather than reduce it directly, e.g. some farmers are planting new heat-tolerant crop varieties.

Carbon offsetting schemes: These allow people to offset their carbon emissions by investing in environmental projects. You can, for example, offset the carbon emissions from a plane trip by paying money into a tree-planting scheme.

International agreements: In 2016, around 200 countries signed the Paris Agreement, an international pact that aims to limit global warming. Today, while many countries are making good progress toward this goal, others have distanced themselves from the treaty.

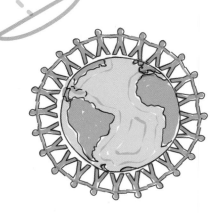

Planting trees: This is one of the easiest and cheapest ways to tackle climate change. Trees absorb carbon dioxide and then lock the carbon away as they grow. Many countries now have ambitious tree-planting schemes.

LIFESTYLE CHOICES

There are many things you can do to help our planet.

Eat less meat: Where possible, meat should be from pasture-fed, free-range, and organic systems. Buying locally sourced, seasonal produce helps to reduce the air miles and carbon emissions produced by food transport.

Waste less: One-third of all the food that is produced in the world goes to waste, while many people go hungry. This has a big environmental cost. About 11% of all the greenhouse gas emissions that come from the food system could be saved if we just stop wasting food.

Travel smart: Walk and cycle short distances. Use public transportation or car shares for longer ones. Take fewer flights, and where you can, use buses and trains to travel.

Chill out: People tend to overestimate the amount of energy used to light their homes, and to heat water. So, turn the thermostat down. And heat your home with energy that comes from renewable sources.

BiODiVERSiTY LOSS AND EXTiNCTION

Biodiversity loss, extinction, and the destruction of nature are each as big a threat to humanity as climate change. Although species go extinct all the time, there are times when extinction rates skyrocket and large numbers of species disappear in very short periods of time. These episodes are called **mass extinctions**. There have been five mass extinctions since life on Earth began, but now scientists think we are living through a sixth period.

Extinction rates are a thousand times greater than during prehuman times. This tells us that human activity is driving the current extinction crisis.

Species at risk of extinction

Human activities are now putting a million species of animals and plants at risk of extinction. Many could disappear within decades.

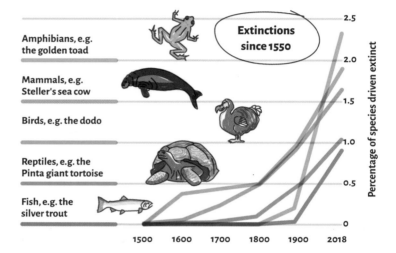

Amphibians, e.g. the golden toad

Mammals, e.g. Steller's sea cow

Birds, e.g. the dodo

Reptiles, e.g. the Pinta giant tortoise

Fish, e.g. the silver trout

Extinctions since 1550

Percentage of species driven extinct

2.5 — 2.0 — 1.5 — 1.0 — 0.5 — 0

1500 1600 1700 1800 1900 2018

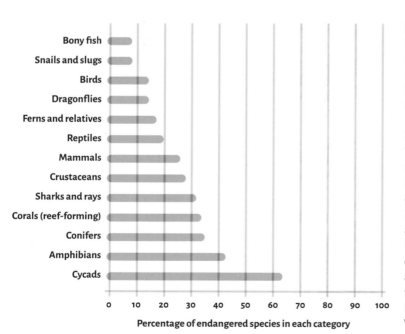

Bony fish
Snails and slugs
Birds
Dragonflies
Ferns and relatives
Reptiles
Mammals
Crustaceans
Sharks and rays
Corals (reef-forming)
Conifers
Amphibians
Cycads

0 10 20 30 40 50 60 70 80 90 100

Percentage of endangered species in each category

The graph on the left shows that more than 40% of amphibians, almost 33% of reef-forming corals, and more than a quarter of mammals face extinction. Invertebrate populations are also declining rapidly.

Earth's biodiversity is crucial. The many services provided by nature to humans are irreplaceable. Nature provides us with food, energy, medicines, and materials. It gives us clean air, fresh water, and soils in which to grow things. It distributes water, regulates the climate, and provides pollination and pest control. As biodiversity falls, we risk losing these services. Humans are a part of nature. We cannot live without it.

Biodiversity threats

Biodiversity is lost when species go extinct, but it is also lost when populations of living things decline. In the last 50 years, populations of mammals, birds, fish, and reptiles have fallen by around 60%. The natural world faces five major threats.

Over-exploitation: The pangolin is covered in scales. In Asia, pangolin meat is eaten and the scales are used in traditional medicine. Poachers take them from the wild illegally, and now the pangolin is the most heavily trafficked mammal in the world.

Habitat destruction: The Tapanuli orangutan was only recognized as a distinct species in 2017. It lives in the Sumatran jungle, but now there are only 800 left. They are threatened by logging, gold mining, and a big new dam that is being planned.

Pollution: The amount of plastic in our oceans has increased tenfold over the last 40 years. This poses a threat to sea birds, turtles, and marine mammals that eat it or become tangled in it. More than half the world's sea turtles have eaten plastic, and more than 100 million marine animals die every year because of this.

Climate change: The Bramble Cay melomys was a small rodent that lived on a single Australian island. Climate change caused the sea level to rise, which caused the island to flood repeatedly. When it went extinct in 2019, it became the first known mammalian extinction caused by climate change.

Invasive species: When non-native, or "invasive," species are introduced to new ecosystems, they sometimes compete with and destroy the native species that live there. The kakapo is a critically endangered species of parrot from New Zealand. Its numbers plummeted after rats, stoats, and other non-native species started eating its eggs and destroying its habitat.

Maintaining ecosystems and biodiversity

As the global population grows, humans are producing more waste and using more resources. This is putting pressure on the environment and its ecosystems. There are many different ways to preserve ecosystems and boost biodiversity.

Breeding programs: These programs aim to boost the numbers of rare and endangered species, e.g. in 1987 the California condor was almost extinct, so all the birds were brought into captivity where they could breed safely. Many of the birds have since been released, and there are now hundreds of individuals in the wild.

Rewilding: This is the simple process of giving land and resources back to nature, so that nature can flourish. This can sometimes involve reintroducing species that have disappeared, e.g. giant tortoises have been put back in the Galápagos Islands where they were once almost extinct. The tortoises have been breeding, and there are now more than 1,500 individuals.

High-tech methods: Conservationists are increasingly using high-tech methods, such as genetics and stem cell biology, e.g. scientists are trying to save the endangered northern white rhino by making test-tube rhinos using frozen sperm and freshly harvested eggs.

Habitat protection: Protected areas, such as national parks and nature reserves, help to preserve ecosystems and help them to regenerate. Marine areas can be protected too. There are more than 13,000 of these. They make up 2% of the world's oceans. Half of these are "no-take" zones, meaning that fishing is forbidden. This allows natural fish stocks to recover.

Ending the illegal wildlife trade:
The illegal wildlife trade is a major
threat to wildlife, e.g. even though
the international ivory trade is
banned, every year 20,000 African
elephants are killed for their tusks.
Governments are making new laws,
and trying to prosecute the criminals
who are caught. Rangers work
hard to protect the animals, and
campaigns try to dissuade people
from buying these products.

Global change: At a global
level, countries need to pull
together to combat climate
change, lessen pollution,
stop destroying habitats, and
produce less waste.

HELPING WILDLIFE

There are many things you
can do to assist wildlife.

★ Rewild an outdoor space by
sowing wildflowers. The blooms
will attract pollinators, such as
bees and butterflies.

★ Be a lazy gardener. Overgrown
flowerbeds, fallen leaves, and
untidy corners provide habitats
for all sorts of creatures.

★ If you have the space, make a
pond. It will attract lots of insects
and other invertebrates.

★ Support the work of charities
and organizations that conserve
wildlife and wild spaces.

Dealing with invasive species:
Invasive species are controlled
with poisons, traps, surveillance,
and detection dogs. In New Zealand,
invasive species kill 25 million birds
a year, so the government has
launched a program—called
Predator Free 2050—that aims
to eliminate all of the country's
invasive vertebrate predators
by that date.

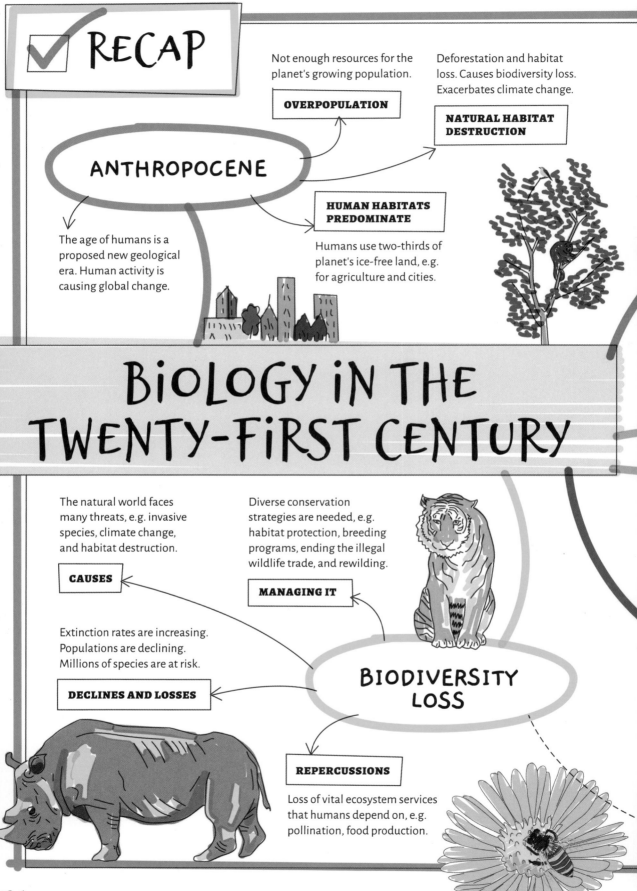

RECAP ✓

Not enough resources for the planet's growing population.

OVERPOPULATION

Deforestation and habitat loss. Causes biodiversity loss. Exacerbates climate change.

NATURAL HABITAT DESTRUCTION

ANTHROPOCENE

HUMAN HABITATS PREDOMINATE

The age of humans is a proposed new geological era. Human activity is causing global change.

Humans use two-thirds of planet's ice-free land, e.g. for agriculture and cities.

BIOLOGY IN THE TWENTY-FIRST CENTURY

The natural world faces many threats, e.g. invasive species, climate change, and habitat destruction.

Diverse conservation strategies are needed, e.g. habitat protection, breeding programs, ending the illegal wildlife trade, and rewilding.

CAUSES

MANAGING IT

Extinction rates are increasing. Populations are declining. Millions of species are at risk.

BIODIVERSITY LOSS

DECLINES AND LOSSES

REPERCUSSIONS

Loss of vital ecosystem services that humans depend on, e.g. pollination, food production.

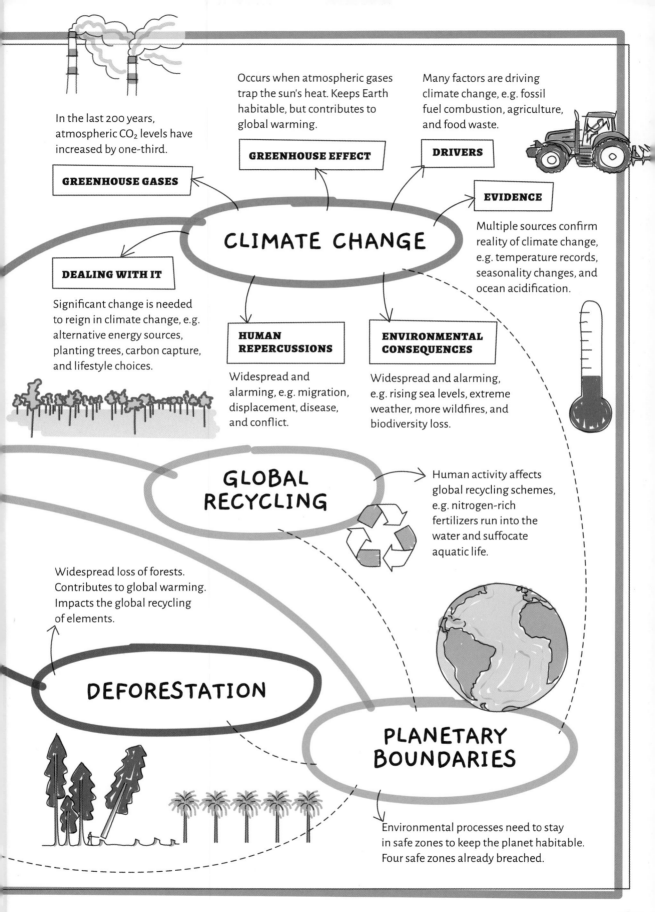

In the last 200 years, atmospheric CO_2 levels have increased by one-third.

GREENHOUSE GASES

Occurs when atmospheric gases trap the sun's heat. Keeps Earth habitable, but contributes to global warming.

GREENHOUSE EFFECT

Many factors are driving climate change, e.g. fossil fuel combustion, agriculture, and food waste.

DRIVERS

EVIDENCE

Multiple sources confirm reality of climate change, e.g. temperature records, seasonality changes, and ocean acidification.

CLIMATE CHANGE

DEALING WITH IT

Significant change is needed to reign in climate change, e.g. alternative energy sources, planting trees, carbon capture, and lifestyle choices.

HUMAN REPERCUSSIONS

Widespread and alarming, e.g. migration, displacement, disease, and conflict.

ENVIRONMENTAL CONSEQUENCES

Widespread and alarming, e.g. rising sea levels, extreme weather, more wildfires, and biodiversity loss.

GLOBAL RECYCLING

Human activity affects global recycling schemes, e.g. nitrogen-rich fertilizers run into the water and suffocate aquatic life.

Widespread loss of forests. Contributes to global warming. Impacts the global recycling of elements.

DEFORESTATION

PLANETARY BOUNDARIES

Environmental processes need to stay in safe zones to keep the planet habitable. Four safe zones already breached.

iNDEX

A

abiotic factors 158, 159, 170
active transport 28, 29, 35
adaptations 56, 87, 160–1, 170–1, 180
adenosine triphosphate (ATP) 98
adipose cells 13
adrenal glands 128, 129
aerobic respiration 98–9, 102, 159, 167
agriculture 49, 63, 179, 180
algae 23, 25, 49, 73, 81
alleles 40–1, 47, 48, 52
alternative energy 180
alveoli 28, 133
amino acids 14, 17, 22, 24, 39, 46, 91, 100
amoeba 19, 25, 73, 76, 89
amphibians 83, 95, 182
anabolic reactions 91, 103
anaerobic respiration 99, 102
angiosperms 81
animals
  cells 13, 22
  competition 162, 163
  drugs from 148
  interdependence 156
  kingdom 72–3, 76, 82–5, 89
  structure and function 116–37
annelids 84
Anthropocene 173, 186
antibiotics 96, 141, 148
  resistance 63, 69
antibodies 14, 143
antigens 143
anus 131
apical meristems 107
appendix 131
arachnids 80, 85
archaea 71, 74, 75, 89
Archaeopteryx 61
arsenic 11
arteries 132, 133, 145
arthropods 84, 85
asexual reproduction 26, 42–3, 52, 80, 85

asymmetrical animals 82
athlete's foot 79, 141
atmospheric aerosols 175
atoms 10–11, 12, 15, 17
atria 95, 132
auditory nerve 126
australopithecines 66, 69
automatic control systems 118, 136
autosomes 38
autotrophs 76
auxins 110–11, 114

B

bacteria 19, 22, 24, 25, 43, 74–5, 89, 96, 164
  antibiotic resistance 63, 69
  classification 71, 73
  communicable diseases 140, 141
  decomposition 167, 169
  denitrifying 169
  extremophiles 161
  nitrogen fixation 169
basal metabolic rate 94
bases 37, 46
behavioral adaptations 160, 170
benign tumors 146
biceps 135
bilateral symmetry 82
bile 131
binary fission 25, 35
binomial system 72
biodiversity 165, 173, 174–5, 179, 182–7
biofuels 49
biomolecules 12–15, 17
biosphere 15, 16
biosynthetic reactions 91, 103
biotic factors 158, 170
birds 45, 61, 62, 64, 83, 95, 155, 163, 183–4
bladder 129
blastocysts 31
blastula 82, 89
blood 29, 33, 95, 129, 131–3, 146
  cells 11, 14, 20, 28, 30–1, 134, 135, 143, 144

glucose levels 118, 128, 136
  vessels 119, 132, 133, 145, 146, 164
Body Mass Index (BMI) 151, 152
body temperature 83, 118, 119, 136
bone marrow 135, 136
bones 134
brain 119, 120–1, 122–3, 129
brain stem 122
breeding programs 184
Broca's area 122
bromeliads 161
bronchi 133

C

calcium 11, 134, 145
calorimeters 94
camouflage 56, 62
cancer 145, 146–7, 153
capillaries 133, 146
capsids 86
carbohydrates 12, 17, 91, 101, 130, 131, 151
carbon 10–11, 12–14, 17, 37, 80
  offsetting 181
carbon capture 180
carbon cycle 101, 166–7, 170, 178
carbon dioxide 28, 33, 98–101, 132–3, 156, 159, 166
  climate change 176–81, 187
  greenhouse effect 176–7
  stomata 109, 161
carbon sinks 173
carcinogens 147
cardiovascular diseases 145, 150, 153
carnivorous plants 80
cartilage 133, 135
catabolic reactions 91, 92, 103
catalysts 92, 102
cells 9, 15, 17, 18–35
  cycle 26
  division 25–7, 31, 35, 42–5
  meiosis 26, 27, 35, 42, 44–5
  membrane 19, 22, 24, 35, 75
  metabolism 90–1

  mitosis 26, 31, 35, 43, 107
  transport 28–9, 35
  walls 23, 24, 35, 113
cellulose 12, 23, 91, 100, 113
central nervous system (CNS) 120, 121, 124, 137
cerebellum 122, 123
chemotherapy 147
chlorophyll 23, 108, 159
chloroplasts 23, 101, 108
chromosomes 26–7, 38–9, 53
  sex 42, 44–5, 52
cilia 76
ciliary muscles 124–5
circulatory system 117, 132–3, 136
classes 72, 73, 83, 85
classification 70–89
climate change 59, 174–5, 176–81, 183, 187
clones 43
cnidarians 84
cochlea 126
cold-blooded organisms 83
combustion 167
common ancestors 55, 57, 64–6, 96
communicable diseases 140–3, 150, 153
community 155, 171
competition 158, 162–3, 171
competitive inhibition 97
complementary base pairing 37
computer modeling 148
concentration gradient 28
condensation 168
cones 124
consumers 156, 171
contraception 143
copper 11
cornea 124–5
coronary heart diseases 145
coronaviruses 87, 88, 142
COVID-19 87, 142–3
CRISPR-Cas9 48, 53
crustaceans 84, 85
culture medium 74–5

cuticle 108, 113, 115
cystic fibrosis 47–8
cytoplasm 19, 22, 24, 25, 35

D

Darwin, Charles 55–7, 61–2, 68
Darwin's finches 55
decomposition 167, 169
de-extinction 49, 53
deforestation 173, 174, 187
deletion mutations 46
denatured enzymes 93
dendrites 30
denitrifying bacteria 169
deoxyribose 37
descent with modification 56
diaphragm 133
differentiation 30, 34
diffusion 28, 35
digestive system 33, 117, 129, 130–1, 136
dinosaurs 59, 61
diploids 42, 52
disease 7, 59, 138–53, 158
    climate change 179
    fungi 79, 113, 141
    plants 112, 113, 114
    see also viruses
DNA 14, 23, 24, 25–6, 36–53
    damage 144, 147
    noncoding 38
domains 71
    three-domain system 71, 88
dominant alleles 40–1, 47, 52
double-blind trials 149
double helix 37, 39, 53
drug development 149, 152
drugs 148–9, 152

E

ears 126–7, 137
    bones 126, 134
ear canal 126
ear drum 126
echinoderms 84
ecology 154–71
ecosystems 15, 16, 80, 155–7, 165, 171, 179, 184–5
effectors 120, 121
eggs 27, 31, 42, 44–5
elbow joint 135
electron microscopes 21, 34
electrons 10
elements 10, 12, 17

embryos 30, 31, 64
endocrine system 117, 128–9, 136
endoplasmic reticulum 22, 35
endothermic reactions 100
enzymes 14, 46, 78, 90–1, 92–3, 102
    digestion 130
    inhibitors 96–7, 102
epidermis 108, 115
epigenetics 7, 51
epithelial cells 32–3
equilibrioception 127
erosion 60
esophagus 130
ethene 111, 114
eukaryotes 22–3, 25, 34, 71, 73, 76–7, 82, 89
Eustachian tube 126
evaporation 168
evolution 54–69
    antibiotic resistance 63, 69
    humans 39, 57, 66–7
    shared features 64–5
    viruses 87, 88
evolutionary trees 65
excretion 9, 16
exoskeleton 84
exothermic reactions 98
extinction 59, 69, 182–3
extremophiles 75, 161, 171
eyes 124–5, 129, 137

F

families 71, 72, 73
far-sightedness 125
feedback inhibition 97
femur 134
fermentation 99
fiber 151
fibula 134
fight or flight response 129
fires 179
fish 45, 83, 95, 158, 159, 164, 183
flagellae 24, 76
Fleming, Alexander 148
flooding 179
flowering 162
food 158
    chains 156, 171
    climate change 179, 180, 181
    competition for 163
    particles 19
    webs 157, 171

fossil fuels 167, 177
fossil record 60–1, 65, 69
fresh water use 175
frontal lobe 122
functional adaptations 160, 171
fungi 25, 73, 76, 78–9, 82, 89
    decomposition 167, 169
    disease 79, 113, 141
    reproduction 42, 43

G

gallbladder 131
gametes 27, 42, 52
gas exchange 109, 133
genes 38–9, 40, 50–1, 53, 57, 144, 147
    editing 48–9, 53
    expression 38
    shared 65
genetic variation 42, 45, 46–7, 56, 58, 59, 69
genetics 36–53, 65, 184
    mutations 46–7, 53, 56, 58, 62, 63, 87
    nature/nurture 50–1
genomes 39, 50, 53, 65
genus 72, 73
geoengineering projects 180
germline cell therapy 48, 53
gibberellins 111, 114
glandular cells 32–3
glucagon 118
glucose 12, 29, 98–9, 100–1, 118, 128, 136
glycolysis 96, 98
Golgi bodies 22, 35
gravitropism 110
greenhouse effect 176–7, 187
greenhouse gases 176–9, 187
growth 9, 16, 26, 43, 79, 98
    plants 101, 107, 110–11, 114, 162
guard cells 109
gymnosperms 81

H

habitat loss/protection 173, 183, 184, 186
haploids 42, 52
health inequalities 139, 152
heart 95, 129, 132
    disease 145, 150, 153
hemoglobin 11, 14, 30
herd immunity 143

hereditary disorders 39, 47, 53
heritability 56, 58, 69
heterogametic sex 44, 45, 53
heterotrophic decomposers 78
heterotrophs 76, 82
heterozygous organisms 40–1, 52
hippocampus 123
homeostasis 97, 118–19, 136
hominins 66–7
Homo erectus 66–7, 68
Homo habilis 66–7, 68
Homo heidelbergensis 67, 68
Homo neanderthalensis 66–7, 68
Homo sapiens 66–7, 68
homogametic sex 44, 45, 53
homologous structures 64
homozygous organisms 40–1, 52
honey bees 85, 164
hormones 14
    human 128–9
    plant 110–11, 114
hosts 164
humans
    evolution 39, 57, 66–7
    health and disease 138–53
    structure and function 116–37
humerus 134
hunger 127
hydrogen 10–11, 14, 17, 37
hygiene 143
hypertonic 29
hyphae 79
hypothalamus 119, 123
hypotonic 29

I

ice cores 178
ice melting 178, 179
identical twins 50
illegal wildlife trade 185
immune system 14, 117, 142–3
incus 126
infectious diseases 140–3, 153
inheritance 40–1, 52
inhibitors 96–7, 102
insects 45, 79, 80, 84–5, 113
insertion mutations 46
insulin 118
integumentary system 117

intercostal muscles 133
interdependence 156–7, 171
interspecific competition 162
intraspecific competition 162
invasive species 59, 183, 185
invertebrates 83, 84–5
iris 124–5
isolation 58
isotonic 29

J

joints 135

K

kakapos 59, 183
keystone species 165, 171
kingdoms 71, 72, 73, 76

L

lactic acid 99
Lamarck, Jean-Baptiste 57, 68
large intestine 131
lateral meristems 107
leaf structure 108–9, 115
lens 124–5
life expectancy 139
lifestyle factors
    144, 145, 147, 153, 181
ligaments 135
light intensity 159
light microscopes 20, 34
lignin 106
Linnaean system 72–3, 88
Linnaeus, Carl 72, 88
lipids 12, 13, 17, 151
liver 99, 118, 131, 144, 151
lock and key model 92, 102
lungs 133, 141, 142
lymphatic system 117
lysosomes 22, 35

M

major elements 11
malaria 76, 139, 140, 141,
    143, 153, 179
malignant tumors 146
malleus 126
mammals 83, 95, 182–3
mass extinctions 182
mechanical papillae 127
medulla 123
meiosis 26, 27, 35, 42, 44–5
Mendel, Gregor 40
meninges 122
meristems 107

messenger RNA 14
metabolic rate 94–5, 103
metabolism 90–103
metamorphosis 84
metastasis 146
microscopes 20–1, 34
migration 160, 179
minerals 9, 29, 100, 105,
    106, 134, 151, 159
    deficiencies 112, 114, 144
mitochondria 22, 30, 31,
    35, 98, 101, 106
mitosis 26, 31, 35, 43, 107
mixotrophs 76
molecules 12–15, 17
mollusks 84
motor cortex 122
motor neurons 120, 121, 137
mouth 127, 130, 137
movement 9, 16
multicellular organisms
    19, 25, 34, 76, 82
muscle cells 30, 32–3
muscular system
    117, 134, 135, 136
mutations 46–7, 53, 56,
    58, 62, 63, 87
mutualism 164, 171
mycelium 79
mycorrhizal networks 79
myriapods 85

N

natural selection
    54, 55–63, 68–9
nature/nurture 50–1
Neanderthals 66–7, 68
near-sightedness 125
negative feedback
    97, 118, 136
negative gravitropism 110
nervous system
    117, 120–1, 129, 137
neurons 30, 120–1, 137
neurotransmitters
    120, 121, 137
neutrons 10
nitrification 169
nitrogen 11, 14, 37
    cycle 169, 170, 174, 187
    fixation 169
nociception 127
noncommunicable diseases
    144–5, 150, 153
noncompetitive
    inhibition 97
non-identical twins 50

nucleic acids 12, 74
    see also DNA; RNA
nucleotides 37, 39, 47
nucleus 19, 22, 23, 35, 38
nutrition 9, 16, 151, 152

O

occipital lobe 122
ocean acidification 175, 176
optic nerve 124
orders 72, 73
organelles
    15, 17, 18, 22, 23, 25
organisms 15, 17, 34
organization
    9, 16, 32–3, 34, 70–89
organs 15, 17, 33, 34,
    117, 124–5, 137
osmosis 28, 29, 35
ovaries 27, 128
oxygen 10–11, 14, 17,
    28, 33, 37, 80, 98–9,
    100–1, 159, 166
    debt 99
    decomposition 167
    metabolic rate 94
ozone depletion 175

P

pain 99, 121, 127, 129, 148
palate 127
palisade mesophyll 108, 115
pancreas 14, 118, 128, 131
parasites 76, 78, 141, 164, 171
parietal lobe 122
parthenogenesis 43
patella 134
pathogens 87, 88,
    140–3, 158
pelvis 134
peppered moths 62, 69
peripheral nervous system
    (PNS) 120, 137
peristalsis 130
pH 93, 159
phenotypes 56
phloem 106, 108, 115
phosphate group 37
phospholipids 13
phosphorus 11, 37
photosynthesis
    28, 80, 81, 90, 91, 100–1,
    103, 109, 156, 159
    bromeliads 161
    carbon cycle 166
    competition 162

phototropism 111
phyla 72, 73, 84
pili 24
pinna 126
pituitary glands 123, 128
placebo 149
plants
    adaptations 161
    cells 23, 29
    competition 162
    defenses 113, 114
    drugs from 148
    growth 101, 107,
        110–11, 114, 162
    interdependence 156
    kingdom 73, 76, 80–1, 89
    respiration 28, 99, 166
    structure and function
        104–15
    tobacco mosaic virus 87
    see also photosynthesis
plasmids 24, 25
Plasmodium 73, 76, 89, 141
pleural membrane 133
polar bears 160
pollination 80
pollution
    46, 51, 59, 62, 175, 183
polymers 37
populations
    15, 16, 155, 171, 186
positive gravitropism 110
predators 156, 158, 171
prefrontal cortex 122
prey animals 156, 171
primary consumers 156
primary growth 107, 114
primary tumors 146
primitive bacteria 75
producers 156, 171
prokaryotes
    24–5, 34, 43, 73, 74–5, 89
proprioception 127
proteins
    12, 14, 17, 22, 39, 46, 151
    see also enzymes
protists 73, 76–7, 89, 141
protons 10
pseudopodia 19
pulmonary circuit 133
Punnett squares
    40, 41, 52
pupils 124–5

Q

quarantine 143

R

radial symmetry 82
radiotherapy 147
radius 134
rain forests 173
randomized trials 149
reactants 91
receptors 120, 121, 137
recessive alleles 40–1, 47, 52
recombination 45, 53
rectum 131
red blood cells 11, 20, 28, 30, 135, 144
red pandas 58
reflex arc 121
reflexes 121, 137
reproduction 9, 16, 42–3, 52
  asexual 26, 42–3, 52, 80, 85
  mate competition 163
  reproductive system 117
  sexual 42–3, 52, 80, 85
  viruses 86, 88
  see also cell division
reptiles 45, 83, 95, 155, 183
resolution 21
respiration 9, 16, 22, 28, 91, 98–9, 102, 167
  aerobic 98–9, 102, 159, 167
  anaerobic 99, 102
  carbon cycle 166
  glycolysis 96, 98
  photosynthesis 101
  plants 28, 99, 166
  water cycle 168
respiratory system 33, 117, 132–3, 136
retina 124–5
rewilding 184, 185
ribosomes 22, 24, 35
risk factors 150–1, 152
RNA 14
rods 124
root hair cells 29
root system 105, 110, 114

S

safe zone boundaries 174–5, 187
saliva 127, 130
saturation 93
scanning electron microscopy (SEM) 21
sclera 124
sea levels 178, 179
sea sponges 19, 82
seasonality 178

secondary consumers 156
secondary growth 107, 114
secondary tumors 146
sediment 60
seed dispersal 162
semicircular canals 126
semipermeable membranes 29
sense organs 124–7, 137
sensitivity 9, 16
sensory neurons 120, 121, 137
sex cells see gametes
sex chromosomes 42, 44–5, 52
sex determination 44–5, 52–3
sexual reproduction 42–3, 52, 80, 85
shared features 64, 69, 72
shivering 119
shoot system 105, 110, 115
sickle cell anemia 47, 144
skeletal system 117, 134, 136
skull 122, 134
slime layer 24
slime molds 76, 77, 89
small intestine 131
social distancing 143
somatic cell therapy 48, 53
somatosensory cortex 122
speciation 58, 69
species 15, 16, 58, 71, 72, 73
sperm 27, 31, 42, 44–5
spider crabs 85
spinal cord 83, 120–1, 123, 137
spine 134
spongy mesophyll 108, 115
spores 27, 77, 79
stapes 126, 134
starch 12, 91, 100
stem cells 30–1, 34, 107, 184
sternum 134
steroids 13
stomach 130
stomata 109, 115, 161
storage molecules 14
structural adaptations 160–1, 170
substitution mutations 46
substrates 91, 92–3
sucrose 12
survival of the fittest 56
suspensory ligaments 124–5
sweating 119

symbiosis 78, 79
symmetry in animals 82
synapses 120, 121, 122, 137
synovial fluid 135
synovial membrane 135
systemic circuit 133
systems 15, 17, 33, 34

T

taste papillae 127, 130
taxonomy 71, 88
temperature 159
  body 83, 118, 119, 136
  decomposition 167
  metabolic rate 94
  optimum 93
  regulation 118, 119
  thermoception 127
  thermoreceptors 119
  see also climate change
temporal lobe 122
tendons 135
territories 163
testes 27, 128
thalamus 123
thermoception 127
thermoreceptors 119
thyroid glands 128
tibia 134
tissues 15, 17, 32, 34
tobacco mosaic virus 87
tongue 127, 137
toxins 96, 113, 148, 162
trace elements 11
trachea 133
transitional fossils 61
translocation 106, 115
transmission of disease 143, 153
transmission electron microscopy (TEM) 21
transpiration 105, 106, 109, 115, 168
transporters 14
triceps 135
true bacteria 74
tuberculosis (TB) 141
tumors 146
twin studies 7, 50–1

U

ulna 134
unicellular organisms 19, 25, 34
urinary system 117

V

vaccines 143
vacuoles 23, 35
vascular system 106, 108, 115
vascular tissue 81
vasoconstriction 119
vasodilation 119
vectors 76, 140, 143
veins 132, 133
ventricles 95, 132–3
Venus flytrap 80
vertebrates 64, 83
vestigial structures 64
viruses 86–7, 88, 142–3
  cancer 147
  communicable diseases 140, 141, 142–3
vitamins 74, 151
  deficiencies 144

W

W chromosomes 45, 52
Wallace, Alfred Russel 55, 68
warm-blooded organisms 83
water 98, 100–1, 159, 166
  cycle 168, 170
  decomposition 167
  fresh water use 175
  nutrition 151
  plant adaptations 161
  transpiration 105, 109, 115
  xylem 105, 106, 115
weather 179
Wernicke's area 122
white blood cells 14, 135, 143
wind 159
Woese, Carl 71
wolves 165
wood wide web 79
woolly mammoths 7, 49, 53

X

X chromosomes 44, 52
xylem 105, 106, 108, 115

Y

Y chromosomes 44, 52
yeast 73, 78, 99

Z

Z chromosomes 45, 52
zoonotic viruses 87, 88
zygotes 31, 42, 77

Acknowledgments

Huge thanks go to Kate Duffy and Lindsey Johns for their fantastic editorial and artistic skills, and to Cynthia Pfirrmann for her expert advice. Thanks also to Joe, Amy, Jess, Sam, and Higgs . . . just for being there.